A Geek Girl's Guide to Electronics and the Internet of Things

A Geek Girl's Guide to Electronics and the Internet of Things

Audrey O'Shea

WILEY

Published by
John Wiley & Sons, Inc.
10475 Crosspoint Boulevard
Indianapolis, IN 46256
www.wiley.com

Copyright © 2021 by John Wiley & Sons, Inc., Indianapolis, Indiana
Published simultaneously in Canada

ISBN: 978-1-119-68368-1
ISBN: 978-1-119-68370-4 (ebk)
ISBN: 978-1-119-68369-8 (ebk)

Manufactured in the United States of America

Library of Congress Control Number: 2020939545

To Jean White, my mom.

About the Author

Audrey O'Shea is a New York State licensed technical instructor teaching electronics and CompTIA certification courses at a technical high school. She also has taught college-level IT courses, holds numerous industry certifications, and has been a mentor to teachers who are new to the profession. Prior to beginning her teaching career, she served as CEO of a consulting company that provided computer training, along with custom software and network installation and support.

O'Shea has degrees in accounting and information technology as well as her teaching license, and she is a member of the Phi Kappa Phi honor society, a public speaker, and a proponent of encouraging women to set their sights on STEAM careers. She has written two books in the self-help genre and served as technical editor on another technology book. In her spare time she enjoys hiking and kayaking in the Adirondack Mountains of New York State, gaming, building and fixing things, and automating her home.

About the Technical Editor

Mike Hitsman is a mechanical engineer working for a small construction company in Central New York. It affords him the luxury to pursue machining, welding, electrical diagnostics, and programming as needed day to day. He attributes his technical breadth to FIRST Robotics, where he started as a student and now mentors the local team. His knowledge of electronics is entirely self-taught, through lots of reading, lots of trial, and lots of errors.

Hitsman lives with Audrey O'Shea's daughter at their country homestead, where he is slowly adding more and more automation to the old farmhouse. He's as likely to be tweaking an IoT device as he is repairing something from 100 years ago.

Acknowledgments

I would like to shower praise on the people who have helped me provide this book for your amusement and education. First, Kenyon Brown for giving me the opportunity. Next, Jan Lynn, my project editor, for pushing me to get it done. Also, a big thank-you to Michael Hitsman, my technical editor, for his polite and concise suggestions on how to make it better. I would be remiss if I failed to mention my friend and fellow author, Mike Meyers, who encouraged me to start writing in the first place, and I would not have made it this far without my mom, who always had my back when the world said, "You can't," and I said, "Yes, I can." Finally, to my family and friends, thank you so much for understanding all the times I said, "I'm sorry, I can't. I have a chapter due." Thank you all.

Contents at a Glance

Contents

Introduction

Welcome to the world of electronics! This is an exciting place to be, and I'm so glad that you've decided to join me in this journey. IoT and electronics are inseparable, as you'll soon see. We'll start with learning about electronic components and creating some useful circuits. In addition to traditional electronic circuits on breadboards, many chapters have circuits that use the programming power of an Arduino board. Finally we'll be connecting your circuits to the Internet so you can control them from anywhere. This book starts small and helps you grow in knowledge as it progresses and encourages you to reach even further. My true hope is that it will get you excited about working with electronics, regardless of whether it's a hobby or a new career, and that it will give you confidence to do whatever it is that makes your heart sing, no matter who or where you are in life.

I'm here to tell you that virtually everyone can do electronics, you included. Although this is a male-dominated field, I encourage you not to let that deter you from getting into the game. If you're female or any other non-traditional geek, follow your passion like I did. Like me, you may be the only woman in the room, but that's perfectly fine. It's an exciting and dynamic career field, and if you're different from the stereotypical electronics geek in some way, that's great! You bring a different perspective, which makes you that much more valuable.

Who Will Benefit Most from This Book

This book is intended for beginners, or those with some basic electronics knowledge who want to fill in those knowledge gaps with tidbits of information that can make all the difference. You'll start with the basics and progress to more interesting (difficult) projects as you go along. This book contains more than 35 "Try This" projects where you can see how things work and solidify what you're learning. Some of the projects include the following:

- Creating a super-bright camping light using LEDs
- Building a laser trip alarm
- Making paper and clay electronics for kids
- Creating light-activated circuits
- Constructing an Arduino wattmeter
- Connecting your projects to the Internet

You'll learn some science behind how things work so the components and connections will make sense, and you'll learn how to avoid some of the pitfalls of circuit building. For example, after studying the material in this book, you'll know why you need a flyback diode for a motor (and understand what one is), instead of having to figure it out when the circuit fails and you're on your third set of components. However, I expect there may be some of that in your future, too, as you take your own ideas and build them into new and interesting circuits. Mistakes are part of the learning process. That's the fun of experimenting! You will create prototype circuits on breadboards, and if you're happy with the result, you can solder them onto something more permanent because you'll know how to do that, too.

Sourcing Parts and Supplemental Materials

I wouldn't want to teach you to fly and then leave you on your own, so throughout the book I've included some of my favorite places to find information and components.

My website, cliffjumpertek.com, also has additional reference materials, tutorials, and videos of the book's projects to help with your journey, should you need it.

Special Features

TIP   Throughout the book, you'll find tips that will provide you with supplemental information, usually to make something easier so you don't have to learn from trial and error.

WARNING   Included in this book are warnings—please take these seriously! Make sure you read all of these because they have information to protect you or others from injury and/or your project from damage.

NOTE   Notes are supplemental material or interesting tidbits of information.

Time to turn the page and get started on your future in electronics! I'll see you there.

What Does This Book Cover?

This book is designed to provide you with a foundation of knowledge in electronics, IoT, and the Arduino environment. Its multitude of projects are the springboard to a deeper understanding and ever more challenging projects.

Chapter 1: IoT and Electronics   The book starts with an explanation of how electronic circuit boards fit into the world of IoT. The components of an IoT system are explained, showing how IoT starts with electronic sensors and then interfaces with computers and communication systems. It also notes some of the challenges in developing and using IoT systems, as well as where IoT may be headed in the future.

Chapter 2: Electricity: Its Good and Bad Behavior   Understanding how electricity behaves will help you to determine why circuits are or are not working. This chapter delves into the science behind electricity. It also explains electricity's characteristics and the relationship between those electrical characteristics. You'll learn how to build a simple circuit and then how to create a circuit on a breadboard, which is the foundation for most of the circuits created in the rest of the book.

Chapter 3: Symbols and Diagrams   The field of electronics uses symbols to represent components when communicating information. This chapter teaches the basic symbology used and how to create a breadboard circuit using a schematic as a guide.

Chapter 4: Introduction to the Arduino Uno   The Arduino is a popular platform for creating IoT implementations. In this chapter, you'll learn the form and function of the Arduino Uno and its programming platform. You'll also learn about analog and digital signals and the role of binary, all while building projects.

Chapter 5: Dim the Lights   This chapter gets its name from one of the eight labs included in it; however, it's more about measuring electricity. You'll learn to use a multimeter and create a voltmeter, ohmmeter, and ammeter using an Arduino board. There's also a great project for creating a camp lamp, and it finishes with a lesson on soldering, perfboards, and shrink tubing, so you'll know how to put those projects together on a more permanent basis.

Chapter 6: Feel the Power   Power is another characteristic of an electrical circuit and is explained in this chapter. You'll learn the relationship between Ohm's law and Watt's law as well as power from batteries, power ratings of resistors, and ways to measure circuit power. You'll also learn how to use an LCD screen with an Arduino board.

Chapter 7: Series and Parallel Circuits   Electricity behaves differently depending on the path that it takes through a circuit. This chapter explains the implications of letting it take those different paths. You'll also learn about the effect of wire size and composition. In the spirit of full disclosure, this chapter involves math.

Chapter 8: Diodes: The One-Way Street Sign   This is the first of several chapters about specific electronic components. You'll learn about diode construction and why diodes behave as they do, along with some common uses for diodes. The projects show you how to work with seven-segment LEDs and bar LEDs.

Chapter 9: Transistors   You may know that transistors are the foundation of our modern computer processors, but they can do so much more. This chapter explains different types of transistors and their uses while giving you some practice working with them.

Chapter 10: Capacitors   Another powerhouse when it comes to computer circuits is the lowly capacitor. They're found everywhere and sometimes taken for granted. This chapter teaches you the characteristics of capacitors and gives you some experience putting them to work.

Chapter 11: The Magic of Magnetism   Magnetism and electricity are like two sides of a coin. This chapter examines that relationship and how magnetism is put to serious work. You'll also learn about relays, which are a component of many industrial electronic circuits.

Chapter 12: Electricity's Changing Forms   Working with electricity is even more fun when you can change it into light and heat and sound, or vice versa. This chapter shows you several ways to do just that, along with the science behind the changes.

Chapter 13: Integrated Circuits and Digital Logic Integrated circuits make our work easier by having an entire circuit on a chip the size of your thumbnail, or even smaller, while digital logic chips can be the decision-makers on our circuits. This chapter explores both and introduces the oscilloscope, which lets you see in real time what is happening electrically in a circuit.

Chapter 14: Pulse Width Modulation Pulse width modulation (PWM) enables a digital signal to control an analog device, such as a motor. In this chapter, you'll build a PWM circuit on a breadboard and then learn how to accomplish the same magic using an Arduino board.

Chapter 15: Sources of Electricity Sources of electricity are touched upon in other chapters, but this one brings all the pieces together in one place. Here you can practice making some electrical current of your own and build an Arduino circuit to monitor the output of a photovoltaic cell.

Chapter 16: Transformers and Power Distribution Transformers are used to change the electrical properties of a circuit and play a big role in power supplies and power distribution. This chapter examines the types and roles of transformers. It also explains how they accomplish changing electrical properties and gives you some experience working with one.

Chapter 17: Inverters and Rectifiers Many times a circuit needs to be converted from DC to AC or AC to DC. The devices that perform that task are called inverters and rectifiers, respectively. In this chapter, you'll learn how both work and how to filter circuits for a more desirable and consistent output. You'll even build a small variable power supply.

Chapter 18: Radio Waves and Tuned Circuits Radio waves are the communication vehicles of our cell phone networks, local area networks, and free music stations. Their use grows every day, so knowing how they work is important whether you're working with computers or IoT devices. This chapter explains AM and FM and introduces working with an Arduino shield to create a radio.

Chapter 19: Connecting Your Circuits to the Cloud Being able to control devices remotely is an important aspect of many IoT implementations. This chapter teaches you how to do just that using a Wi-Fi enabled Arduino board and the Arduino IoT cloud.

Chapter 20: Just for Fun While working with electronics is always fun, most of the time electronic circuits are created for serious work. This chapter explores some of the not-so-serious uses of electricity. If you're the least bit creative, and I'm sure you are, then this chapter is for you!

IoT and Electricity Basics

In This Part

IoT and Electronics

"Toto, I have a feeling we're not in Kansas anymore."
—Dorothy, *The Wizard of Oz*

Sci-fi movies and shows have always been my obsession. As a child, I would watch in awe when reruns of *The Jetsons* showed people talking on video phones, the almost-human robot maid, and sidewalks that moved. Then there were the *Star Trek* reruns where characters walked around with communicators that allowed them to talk to anyone just by tapping them. Later, *Star Wars* had language translators that would automatically translate into any other language . . . rather like what Google Translator does now.

Forty or fifty years ago, many devices and capabilities that are commonplace now were considered ridiculous, impossible, or mere fantasy. Were these movies and TV shows predictions of the future, or did they help to shape the future by putting these notions of "impossible" devices into someone's mind to start working on? Either way, many of those devices exist now for us in some form or another. Even 20 years ago, most people still depended on their home phones for communication. Do you know anyone who still has a landline at home? They are few and far between.

The Electronics Technicians Association was founded in 1978 as the electronics industry was beginning to grow slowly. Now, it's growing by leaps and bounds on a daily basis. It's astonishing how far electronics have come in such a relatively short time when compared to human existence, and it's even more incredible when we ponder how far we will be 50 years from now. Many of the technological advances of the future will be here due to artificial intelligence, machine learning, and the myriad of sensors starting to cover our world.

The world is about to take another leap forward, and if you want to be part of that journey, learning electronics is the place to start. As so many maps show us . . . "You Are Here."

IoT in a Nutshell

What is IoT? As you may know, IoT stands for "Internet of Things." IoT refers to a vast array of connected devices that gather and transmit data over interconnected networks with or without human intervention, sometimes even responding to the captured data automatically as machines talk to machines and learn from each other. (When IoT involves manufacturing processes, it is often called industrial IoT [IIoT].) IoT can include data gathered by proximity sensors on your car's front that detect deer in the roadway and signal your brakes to immediately slow down the car without you doing anything. It also includes when moisture levels (or lack thereof) are transmitted from a field to that field's watering system, signaling to turn on the irrigation system without a human lifting a finger. Even a dog's GPS-enabled location device is part of an IoT system, as is the Tile that I press to locate my often-misplaced car keys.

Other systems considered part of IoT are smart cities, smart grids, smart homes, smart watches, and manufacturing machines talking to and learning from each other. Smart devices are used in hospitals, schools, retail, and nearly any other service or business you can think of. Last year, I attended a virtual meeting with someone from a major networking device company. He was speaking from his office about power over Ethernet and how the interconnected devices controlling heat, lighting, air quality, etc., all ran automatically in the high-rise office building he was in. I noticed a model of a pig on the credenza behind him and asked about it. Yes, it was a pig wearing an IoT collar.

What does this have to do with learning electronics? Everything! Electronic sensors and circuits are the beating heart of an IoT system.

Parts of an IoT System

What comprises an IoT system changes depending on who you ask, but regardless of what particular twist an industry or company may put on it, certain things must be there. For an overview of an IoT system, see Figure 1.1.

Devices

What is an IoT device? IoT devices include sensors, circuits, software, actuators (things that do something, such as switch from one state to another), and microprocessors, all rolled into a neat little package. These devices also need a way to communicate and send data to a place where it will be processed, manipulated, and action taken based on the data, or they need to be able to communicate to receive instructions based on the data that was gathered by some other device. Therefore, an IoT device can be on either the sending end or the receiving end, or possibly both.

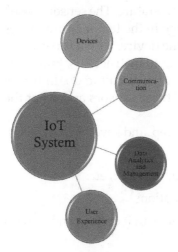

Figure 1.1: An IoT system

Take, for example, a smart home with a remote-controlled thermostat, which has a few layers of things going on. First, the thermostat is a device. It has a sensor that measures the temperature and sends that information to a circuit board with a microprocessor where the reading is converted into data, which is manipulated. If certain conditions within the software program are met, the microprocessor sends a command to another component, telling it to turn the furnace either off or on. This example is machine to machine but involves sensors, circuits, software, microprocessors, communication, and actuators.

Another function of this device would be the ability to access the device from a cell phone via the Internet and Wi-Fi to tell the device to turn up the heat before the user gets home. This example involves a user interface, which is part of the entire user experience, but here we have the cell phone acting as a device and the thermostat acting as another device, communicating via the Internet.

Sensors

The first part of any IoT system is a device that senses something physical, whether a particular condition or event. It could be a fiber optic cable in a building's concrete that picks up a pressure change, or it could be a proximity sensor, heat sensor, humidity sensor, optical sensor (ambient light, IR, UV), gas sensor, position sensor, magnetic sensor, motion sensor (accelerometer, gyroscope), color sensor (light again), or touch sensor (pressure). A search for *sensors* on an electronics components site at the time of this writing yielded tens of thousands of results.

As mentioned in the preceding section, a sensor receives some form of raw data and passes it on to a circuit and most likely a microprocessor, where the data received is interpreted. Take, for example, a temperature. The sensor doesn't send "temperature" to the circuit. Instead, a change in the temperature causes a rise or fall in either current or voltage through the device, which is passed to a circuit where it is read and interpreted according to instructions, known as *software*, controlling what the microprocessor tells the circuit to do. In future chapters, you'll learn how to work with some of these sensors and what the technology is that drives them.

Choosing the right sensor is an important first step, and several characteristics need to be considered.

- **What's being measured:** Temperature, light, pressure, etc.
- **Electrical:** Current, voltage, and power limitations.
- **Physical:** How much pressure, heat, light, etc., can it endure and remain viable?
- **Accuracy:** How far might it vary from the actual measurement?
- **Sensitivity:** How much does the input need to change before the output changes?
- **Reliability:** What is the track record of this sensor? When does it stop being accurate? How often does it break down?
- **Range:** What are the minimum and maximum values that can be measured?

Circuits, Software, and Microprocessors

Once the right sensor is found, the circuitry and software can be designed to interpret the information that is provided by the sensor. A myriad of choices exists for all of these. A microprocessor can be a single chip designed to perform logic or a microprocessor platform, such as an Arduino device. The choice of processor may determine the choice of software that is used.

Communication

Because the "I" in IoT stands for Internet, the assumption can be made that the device is expected to be interconnected in some way either through a local area network (LAN) or through the Internet. However, different levels and aspects of communication may be used by IoT devices.

Levels

A device may communicate with other devices, such as a moisture sensor in a field that triggers a watering system to work, or devices in a factory assembly line that communicate with other devices to either slow down or speed up the processes based on conditions. This is *machine-to-machine communication.* Systems like these may even use *machine learning*, which is a process where computers use algorithms to look for related data and learn, changing their programming based on data without human intervention. Machine learning is far too complicated to explain here but definitely an emerging technology worth paying attention to.

In the automatic watering system example, the devices may be connected only to a LAN, but most likely they will be gathering data and sending it, via one or more protocols and networks, to a place where it is processed.

Protocols and Standards

Protocols are essentially rules for communication, and they are the topic of much learning and discussion in the computer world. *Standards* typically define how a network is built and what protocols are used on it. Networking standards are needed to ensure that different devices, possibly from different vendors or manufacturers, are all able to communicate effectively. Knowing what type of communication is needed and used by each part of an IoT system is an important consideration. Any of the following standards might be part of an IoT system:

- **Ethernet:** Wired LAN networking standard
- **Bluetooth:** Short-range wireless, usually connecting mobile devices
- **Wi-Fi:** Wireless networking standard allowing wireless networks to connect to a wired system or each other
- **Cellular:** For longer-range wireless connectivity via the cellular system

Each of these standards may have multiple protocols that are used with that particular standard, and any of them may allow a device to connect to the Internet. Often more than one will be used. In the case of the smart thermostat, it may use Wi-Fi to connect to a local LAN using Ethernet, which is connected to the Internet via some other method, such as a cable modem, fiber optics, or satellite. The remote user may be connecting their cell phone to the thermostat via Wi-Fi or the cellular network. All of these standards need to work in harmony for successful communications across an IoT system.

Data Analytics and Management

Data analytics is the growing field of sorting and analyzing raw data to derive meaningful and actionable information from it. It is the stuff of algorithms and perhaps insight gained from the experience of working with data. Four basic types of data analysis exist.

- Descriptive
- Diagnostic
- Predictive
- Prescriptive

Descriptive analytics identifies what has happened based on data. It is often based on key performance indicators (KPIs). For example, did sales go up or down and by how much or what percentage? Did the field have to be watered more often or less often?

Diagnostic analytics attempts to explain why the changes occurred. It can look at data outliers and what was occurring when the change in data occurred.

Once the what and why are known, *predictive data analysis models* can be built to anticipate problems before they happen or identify future trends.

Finally, *prescriptive data analysis* gives the user a course of action to take to avoid the problems identified in predictive models or to take full advantage of what's around the corner.

Data is more valuable than the hardware and software used to mine it from various sources, so managing that data is a primary concern for any business. The field of *data management* is concerned with collecting data, maintaining its physical security, securing the privacy of any personally identifiable information (PII), and preserving that data in a cost-effective and efficient manner. Data analytics is part of data management, and the way this data is used can cause a business to flourish or fail.

The User Experience

The *user experience* refers to every place that the user and the IoT system come into contact with each other. If, for example, an employee is tasked with identifying a problem quickly and accurately, then the interface that the employee uses needs to have the right information in an application that is easy to use, and it needs to be updated in a timely manner.

Depending on the situation or business, the user experience can also include interacting with customer service or other personnel, how easy the phone system is to use, etc.

Challenges in Implementing IoT

Implementing technology is not without challenges, particularly in an IoT system. One of the biggest challenges for remote IoT devices is power. Running electrical wires is not always practical, and batteries have a limit to their capacity, so creating a system that runs on as little power as possible, or perhaps renewables like solar power, is a significant hurdle to overcome in developing an IoT solution.

Another major concern is the security and ownership of data. If data is stored by a provider, who owns and has access to the data? What encryption will be used to transmit the data from where it's gathered to where it's used? These are questions that any businessperson would want to know before implementing a system. Determining the right data to gather is probably the first question to answer, because measuring the wrong facts won't help a business make the right decisions.

Perhaps the biggest challenge of all is cohesiveness. If a system is cohesive, then it works together smoothly, which can be a problem when so many different systems are involved. Beginning with the sensors that detect data, through the circuitry and networks that transfer and store the data, then send the data to a user's interface on a device such as a phone or computer, and back again as the user responds through a local network to the web and then to the actuator; during this process, there needs to be a seamless way to transmit and manipulate the data. With so many protocols and systems involved, that can be difficult indeed. Developing a system needs to start with a bird's-eye view of the major parts, working down to the component level to ensure that all of the system's parts work together to move data around smoothly.

IoT into the Future

IoT and IIoT will not be going away in the foreseeable future. In fact, they will continue to grow as processes and machines get smarter and people find more uses for IoT. What may have started as a curiosity, with devices that were more fun than function in the hands of a few hackers, is now a major industry and will be causing paradigm shifts in virtually all industries and businesses.

Devices and sensors will continue to get smaller, get less expensive, and work better. We will learn how to better harvest the data from IoT devices and put them to more and more uses. Already billions of IoT devices are being used, and the number increases exponentially as every day forward-thinking inventors and electronics enthusiasts devise more uses for the technology. Where will you fit into this growing industry? Perhaps the best place to start is with a basic understanding of electronics, and to that end, read on.

CHAPTER

2

Electricity: Its Good and Bad Behavior

"Never trust an atom. They make up everything."

Unknown

You probably already have an idea of how electricity behaves. If you turn on a light switch, electricity is converted to light. It makes motors run and can be created by a generator to charge our cell phones when the power is out. You also know that it's a bad idea to stand outside in a lightning storm because it's likely also raining, and if you're wet and the tallest thing around, you're practically inviting lightning to strike you. Electricity can keep us alive if it's powering our heart, and it can kill us if we don't respect it. What you may not understand is *why* electricity behaves the way it does, which is what this chapter is about. If you're going to be the next Bill Gates or Steve Jobs and invent something that will alter life as we know it, you'll need to start with a basic understanding of how and why electricity behaves the way it does.

First, a little static electricity lab.

Try This: Creating Some Static

This lab is just for fun. Most grade-school kids have rubbed a balloon on their head or combed their hair with a plastic comb and seen the magic that is *static electricity*. What happens is that the energy of friction pulls electrons from the atoms of the hair onto the comb or balloon, giving it a negative charge and the hair a positive charge. The negatively charged balloon attracts the more positively charged hair. (Opposites attract.) If two balloons are negatively charged, they will push away from each other. (Like charges repel.) You can also do other fun things with static. The following sections contain a few for your amusement.

For these projects, you need the following materials:

- Fur, wool cloth, or hair (a source of electrons)
- PVC pipe about 2' long
- Styrofoam plates
- Water faucet
- Balloon
- Fluorescent tube
- Water glass or glass vase
- Styrofoam ball
- Aluminum foil
- String

Levitate a Styrofoam Plate

1. Rub a Styrofoam plate with a wool cloth to charge it and then set it on a flat surface.

2. Rub a second Styrofoam plate to charge it as well.

3. Try to put the two plates together. If they push away from each other, you know they're both negatively charged. Now, put your hand a few inches above the plate that's on the table and try to place the other plate on top of the first one. It should float up to your neutral (no charge) hand. It is being pulled by your hand and pushed by the other plate.

Bend Water

1. Negatively charge the PVC pipe by rubbing it with the fur or wool cloth.

2. Turn a water faucet on so that it is running a small steady stream.

3. Move the charged PVC pipe near but not touching the water. You should see the water bend toward the more negatively charged PVC pipe (Figure 2.1).

Creating Light with Static

1. Negatively charge the balloon by rubbing it with the wool cloth, hair, or fur.

2. Enter a darkened room.

3. Touch the balloon to the two electrodes sticking out of the fluorescent tube. The tube completes the circuit and inside the tube the electrons excite the gasses and cause the glow.

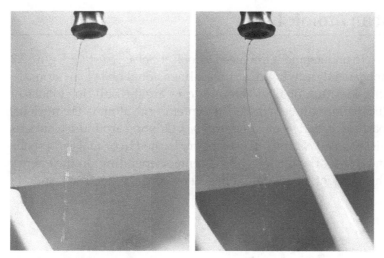

Figure 2.1: Bending water

The static electricity generated this way will have enough voltage but not enough current to light a light-emitting diode (LED), which we'll talk about later. It *does* have enough voltage to excite the electrons of the gasses in the fluorescent tube. The process is similar to someone walking across a carpet and then touching a metal doorknob. They may get quite a shock when the negatively charged electrons jump to the more positively charged doorknob. The electrons move quickly and may cause sound or a flash of light and be thousands of volts but not a lot of current. Current and voltage will be explained soon.

Magically Move a Styrofoam Ball

1. Wrap a small Styrofoam ball with aluminum foil.

2. Tie a string around the ball.

3. Tape the end of the string to the inside bottom of the water glass or glass vase, making sure that the ball will hang freely when the glass is turned upside down.

4. Turn the glass upside down.

5. Charge the PVC pipe and bring it toward the glass. The ball is attracted to the negatively charged PVC pipe.

While these were fun demonstrations of static electricity and the *law of charges* (like charges repel each other, opposite charges attract), static electricity does have some serious industrial uses. Static electricity is responsible for transporting toner inside a printer from the negative toner container to the more positive (but still negative) drum and finally onto the positively charged paper. Static is also used in some pollution control systems where particles are charged and then attracted to plates with the opposite charge, reducing pollution. Static is also used in applying paint to cars. What causes those charges? Read on.

Electricity at an Atomic Level

What is electricity? To understand it, you need to look at *atomic structure*. Figure 2.2 shows a two-dimensional drawing of a three-dimensional object, an atom. To be specific, it is the structure of a gold atom. Atoms are the building blocks of everything, including human beings. At the center of the atom is the nucleus, which contains particles called *protons* and *neutrons*. Orbiting around the nucleus is a cloud of particles called *electrons*. Electrons are located in orbitals and shells at various distances from the nucleus in the center depending on the energy they exhibit at the moment. A gold atom has 79 protons, 79 electrons, and 118 neutrons.

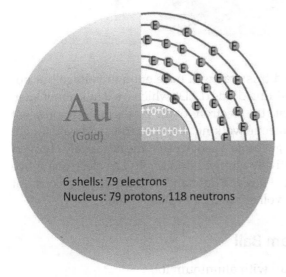

6 shells: 79 electrons
Nucleus: 79 protons, 118 neutrons

Figure 2.2: Atomic structure of gold

Matter that is made of only one type of atom is called an *element*. These elements and the information about each can be found on the periodic table of elements (Figure 2.3). While there are 94 naturally occurring elements, more have been created by humans. Each element is assigned an atomic number, which is equal to the number of protons in the nucleus of the atom, so our gold atom's atomic number is 79.

Protons have a positive charge, neutrons have no charge, and electrons have a negative charge. Atoms seek to be in balance, so when they are at a ground state, atoms always have the same number of protons and electrons, making the atom have a net neutral (no) charge. When an atom is acted upon by some outside force such as friction, it can lose or gain electrons. This process of losing and gaining electrons is called *ionization*. Because electrons are negative, if the atom loses an electron, it becomes a *positive ion*. It will have more positively charged particles (protons) than negatively charged particles (electrons) and therefore a net positive charge. If an atom gains an electron, it becomes a *negative ion* because there are more negatively charged particles (electrons) than positively charged particles (protons).

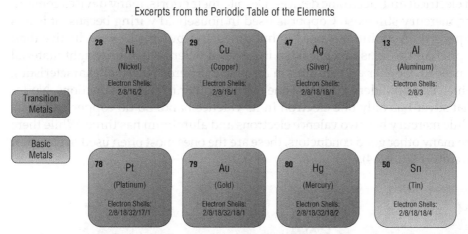

Excerpts from the Periodic Table of the Elements

Figure 2.3: Excerpt from the periodic table of elements

NOTE   The designations of positive and negative were chosen by Benjamin Franklin as a way to explain his observations of electrical behavior.

The law of charges tells us that opposites attract, so the positively charged protons are always pulling on (attracting) the negatively charged electrons. When ionization occurs, the electrons the atom loses (or gains) will be located in the outermost shell, which is called the *valence shell*. If an electron in a lower shell is acted upon by some outside energy, such as heat or light, it can jump to a higher shell. When it loses its energy, it falls back down toward the nucleus. Only a certain number of electrons can exist in a given shell, but right now we don't need to explore that any further.

What does this have to do with electricity? Everything! What we know as electricity is the movement of those electrons from atom to atom in the same general direction, as they're trying to balance atoms in a chain reaction.

Conductors and Insulators

Certain elements will easily give up their valence electrons. We call those elements good *conductors*. Examples of good conductors are gold, aluminum, copper, silver, and mercury. Each of these conductors has characteristics that make them better than the others in certain situations. For example, silver is the best conductor, but gold is often used for connections on computer boards because of its tendency to avoid corrosion. Mercury is a liquid that has been used in devices such as thermostats, thermometers, and motion switches, and for measuring pressure. However, because it has been identified as a pollutant, the electrical and electronics industries have been working to replace mercury in electrical and electronic devices. Despite their efforts, many devices containing mercury still exist. Copper is used in household wiring because it is less expensive. Aluminum is used in buildings, too, but it is less conductive than copper and weighs much less, so it is useful where a lighter-weight material is needed. Copper and aluminum also have different thermal characteristics, which leads builders to choose one over the other in certain situations. Notice that silver has only one electron in its valence band. So do copper and gold, while mercury has two valence electrons and aluminum has three. While there are many other good conductors, these are the ones most often used in electrical circuits (Figure 2.4).

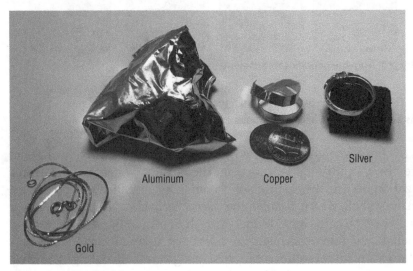

Figure 2.4: Good conductors

Materials that are good *insulators* are glass, plastic, rubber, and dry wood. Insulators do not readily give up electrons, and most are compounds, meaning that they are made from more than one type of atom. Rubber, for example, has the chemical composition C_5H_8, which is five atoms of carbon and eight atoms of hydrogen. Glass is made from silicon and oxygen. Other materials, such as boron and cobalt, are added to glass to change its properties. Boron, chlorine, and sulfur are elements that are considered insulators. Sulfur has six valence electrons and chlorine has seven, while boron has only three.

Silicon and germanium are *semiconductors*. In pure form, they are not good conductors or insulators. Yet, when their properties are changed by doping them with other chemicals, they become useful. Both are used in electronic circuits. Some diodes, which are explained in Chapter 8, "Diodes: The One-Way Street Sign," are made of germanium, while silicon is the building material of integrated circuit chips. Silicon and germanium both have four valence electrons.

A single element can exist in different forms called *allotropes*. Allotropes of the same atom may behave differently. For example, carbon in the form of graphite is a conductor, but when compressed over time into a diamond, it acts as an insulator (see Figure 2.5). When in doubt, do your research to confirm a material's properties before you use it in your circuits.

Figure 2.5: Allotropes of carbon

Human beings, by the way, can be good conductors. Our skin is a decent insulator, so very low voltages are safe to handle. However, higher voltages, such as the 120VAC found in U.S. household wiring, can prove fatal or at the least provide an uncomfortable shock. Wet or sweat-soaked skin becomes more conductive, so be sure to use caution when working with electricity.

Characteristics of Electricity

Electricity has certain characteristics that you must understand to determine if the circuit you want to build will work. The first three characteristics to be aware of are *current*, *voltage*, and *resistance*. Current, voltage, and resistance must exist in a particular relationship to each other for a circuit to work as intended or sometimes to work at all. If they are not in balance, the circuit may still work, such as a light will light, but without enough resistance damage will result. If there is too much resistance on the circuit, the light won't turn on. Electricity flowing through a conductor is similar to water flowing through a hose, so we'll use that analogy to help explain the relationship.

Current

Standard units are needed to make comparisons and calculations in a circuit. The volume of electrons through a conductor is called *current*. Current is measured in *amperes*. Electrons are extremely tiny. It takes 6.24×10^{18} electrons to make 1 *coulomb*, and 1 coulomb of electrons passing a point in 1 second is an ampere of electricity (Figure 2.6). Amperes, which are coulombs per second in a circuit, are akin to gallons per hour (GPH) through a water pump. Considering this, at any given second, if there is 1 ampere of current in a circuit, there are 6,240,000,000,000,000,000 atoms whose electrons are bouncing to the next atom and displacing its electrons at any given point in that circuit. I is the symbol for current, so as not to confuse it with coulombs, whose symbol is C. The symbol for an ampere is A.

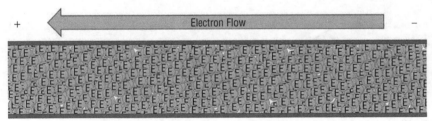

Electrons bounce from one atom to another inside a conductor.

Figure 2.6: Electrons flowing in a conductor

Voltage

Think now of a water hose. When it's turned on at the faucet, you can feel pressure pushing water through the hose. This pressure is like the electrical characteristic called *electromotive force* (EMF). EMF is also known as *potential difference*, and it's a measurement of the difference in electrical pressure between two points. EMF is measured in *volts* and is often used interchangeably with the term *voltage* even though that's not exactly correct. For example, a battery may have a potential difference between its positive and negative terminals of 9 volts. When those two terminals are connected by a conductor, the electrons from the negative terminal rush to fill the "holes" in the atoms at the positive terminal, but in their disconnected state the difference is 9 volts. A volt is defined as the amount of electrical pressure needed to push 1 ampere of current through 1 ohm of resistance. E is the symbol for EMF, and V is the symbol for volts.

WARNING   Safety Tip! Never connect the negative and positive terminals of a battery directly together without a load. Doing so will create excessive current. The conductors can overheat and cause a fire, or the battery may get very hot, swell, or even explode. Don't do it!

Resistance

Think again of the water hose. If you grab it with both hands and kink it, the kink in the hose is *resistance*. When you kink tightly, no water can flow. If you let the kink relax a bit, some of the water will flow through but not all. The kink in the hose is limiting the water flowing through the hose, just like resistance in a circuit *limits the current* flowing through the circuit. We will review specific devices called *resistors* later, but for now understand that any object in a circuit can provide resistance, whether it's a lamp, a diode, or a speaker. Even the conductors themselves will have some resistance. Resistance is measured in a unit called the *ohm*. An ohm of resistance exists in a conductor that will allow 1 ampere of current to pass with a pressure of 1 volt. R is the symbol for resistance, and the Greek letter omega (Ω) is the symbol for ohm.

CHARACTERISTIC	DESCRIPTION	SYMBOL	MEASURED IN	SYMBOL
Current	Number of electrons	I	Amperes (amps)	A
Electromotive Force	Electrical pressure	E	Volts	V
Resistance	Limits current	R	Ohms	Ω

Induction and Conduction

It's important to understand induction and conduction. Induction and conduction occur because of the law of charges. *Conduction* happens when something charged comes into contact with something neutral or of a different charge (Figure 2.7). The electrons in the more negatively charged object will rush into the other object, so they both take on the same charge. This is the normal flow of electrons from a source through a conductor.

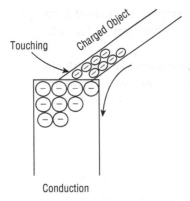

Conduction

Figure 2.7: Conduction

Induction happens when something with a charge comes close to but doesn't touch another object (Figure 2.8). If the charge is negative, the extra electrons in the charged object will push away the electrons in the object close to it (remember, like charges repel each other), causing a positive charge in the second object. If the charge in the first object is positive, the electrons in the uncharged object will be attracted to it, causing the uncharged object to take on a negative charge. Regardless of the polarity of the charge, induction causes the uncharged object to take on the opposite charge of the first one.

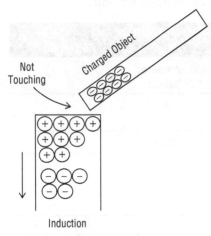

Induction

Figure 2.8: Induction

WARNING   Safety Tip! Remember, atoms always seek to be in balance and opposites attract, so a negative or positive charge is always trying to move to something more neutral. The ground will always be more neutral than an electron, so unpaired electrons will always go toward ground. This is why household wiring is literally grounded with metal rods driven into the ground to give electricity a safe path to follow to ground. This grounding can prevent house fires and helps prevent people from getting shocked.

Ground isn't necessarily "the" ground. A ground in electrical terms can be the more negative side of the circuit on which other points are based or against which measurements are taken, or it may mean a literal connection between a circuit and the earth (i.e., a path specifically created for electrons to safely be united with atoms).

Electricity is also lazy. It will take the easiest path to ground. Don't inadvertently become that path! Normally the power is off when you're touching a circuit, but be aware of your surroundings. Wear shoes with nonconductive soles (like rubber) to put a barrier between you and the ground. Remove metallic jewelry while working with electricity. Look around and be aware of conductive objects near you such as metal flashing, a metal sink, or other conductive materials that might come into contact with what you're working on. As little as 10mA (.01A) can cause pain, and 100–200mA (.100–.200 A) of current can cause death, but the amount of damage it does depends on how long the current flows through a person and the path it takes.

Try This: Creating a Simple Breadboard Circuit

An *electrical circuit* needs a complete path from the source of electrons (negative) to something more positive, such as from a battery's negative terminal back to its positive terminal, through a load of some sort like a light or a fan, which converts the electrical energy into another form of energy, in this case light or movement of air. (Incidentally, those devices that convert between electricity and some other form of energy are called *transducers*.) Conductors provide the path from negative to positive. A simple circuit could be a 6V lantern battery connected with alligator clips and conductors to a 6V lamp and back to the battery. An alligator clip has jaws that open and close around an object to hold it tightly. Alligator clips can be purchased with conductors attached. The conductors are wires made of metal coated with insulating material to keep the electrons on the path that you want them to take. See Figure 2.9.

Another way to connect circuits is through soldering (Figure 2.10). Soldering involves using a hot iron to melt a metallic alloy, which electrically joins the metal leads of two components together, usually via the traces (conductive tracks) on a printed circuit board (PCB). Component leads can also be directly soldered together without a board if the project is small.

While using alligator clips is simple, it would be cumbersome to build complex circuits that way. Soldering circuits is great for the final product, but not so great when you're still working on it and tweaking it to be better. Thus, enter the *breadboard* (Figure 2.11). Breadboards make life much easier and are a great place for novices to start putting circuits together. A breadboard provides a way to *temporarily connect* circuits without having to solder or clip wires together.

However, the biggest advantage is that you can change your mind, move things around, try different components, and recover more easily from mistakes than when you're soldering. If you run out of room for complex circuits, simply add another breadboard to expand your workspace. These tiny breadboards (Figure 2.11) have clips on the ends to make connecting them simple and more

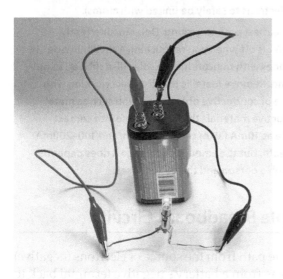

Figure 2.9: Simple electrical circuit

Figure 2.10: Soldering on a PCB

secure.

A breadboard has holes for inserting components and jumper wires, with metal strips underneath that making connections for you (Figure 2.12). The other side of the metal rows has clips that hug the metal leads of components

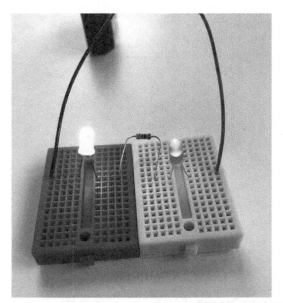

Figure 2.11: Expanding breadboards

Figure 2.12: Bottom of breadboard

and wires that are inserted through the holes (Figure 2.13).

The long strips on the edges provide a way to get power to both ends of the board. They are called *power rails* (Figure 2.14). A long metal connector runs underneath them. Power can be connected simply by inserting wires from a

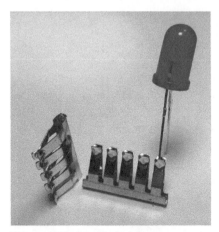

Figure 2.13: Breadboard clip

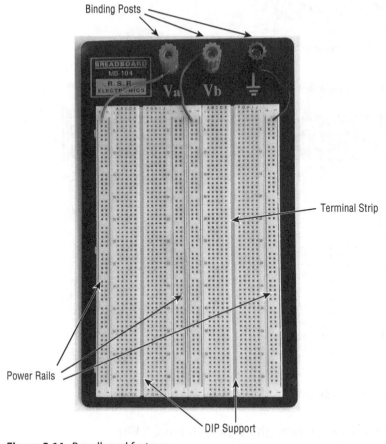

Figure 2.14: Breadboard features

source into the power rails.

Your breadboard may (or may not) have binding posts (Figure 2.15). These posts enable you to connect power to the breadboard from a power supply unit (PSU) using banana clips. Each binding post will have a hole in it where a stripped wire is inserted and then the post tightened down onto the wire to make the electrical connection from the binding post to the breadboard. The other end of the wire is plugged into a power rail.

An advantage of using a PSU is that the voltage or current applied to the circuit can be changed quickly by simply turning a knob or pushing a button, depending on the PSU being used. The power can also be disconnected quickly by turning off the PSU rather than unplugging a wire.

If a breadboard has positive and negative power rails on both sides, a good practice is to jumper between the two sides so positive and negative power connections are available on both sides of the breadboard. Red is for the positive side of the circuit, and usually a blue or black stripe is used for the negative side

Figure 2.15: Binding post close-up

of the circuit (Figure 2.16).

Jumpering the rails together this way is helpful when wiring complex circuits because there is no need for long jumper wires connected from one side of a component to the opposite side of the board. Simply jumper between the component and the closest power rail. Following the red and blue/black color standard will help you and anyone who is looking at your circuit easily see how components are connected.

Terminal strips are the shorter rows of usually five or six holes that can be used to connect components. Any lead inserted into the same row will be elec-

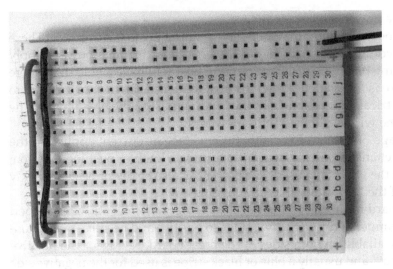

Figure 2.16: Power rails jumpered together

trically connected to the other devices in that row.

The dip in the middle is for, well, DIPs. DIP stands for dual in-line package, and there are many integrated circuit (IC) chips that are in DIP format, made to fit into a breadboard. The dip in the breadboard creates an electrical separation between the two sides of an IC DIP so each pin can be connected independently

Figure 2.17: An IC in DIP format across a breadboard dip

(Figure 2.17).

Notice that the rows of the breadboard are numbered, and the columns are notated with letters. These marks are helpful in determining where to connect components and in explaining to someone else how components are connected. You could say, "The lead is in row 30, column D," instead of simply stating "It's

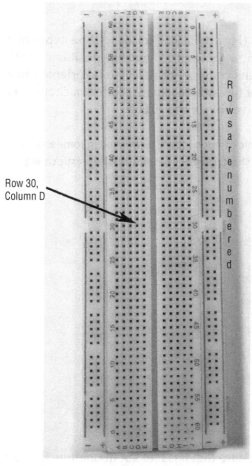

Row 30, Column D

Columns Use Letters

Figure 2.18: Breadboard rows and columns

about in the middle of the board" (Figure 2.18).

It's time to create a simple circuit on this breadboard.

For this project, you need the following materials:

- Breadboard
- (3) LEDs—most voltages will work, but higher-voltage LEDs (white or blue) will work best

- 22-gauge wire—just a few inches—or premade jumpers
- Wire strippers (if not using premade jumpers)
- 9V battery
- Battery snap

Light-Emitting Diodes

LEDs can be both practical and fun (Figure 2.19). Diodes will be explained more later, but for now you need to know that an LED must be positioned with proper *polarity*. Polarity means that the component needs to be oriented in a specific way with a negative connector toward the negative side of a circuit and a positive connector toward the positive side of the circuit.

NOTE The negative side of an electronic component will be marked in some way. In the case of an LED, it's usually noted by a shorter leg or a flat side on the plastic casing of the LED.

Figure 2.19: LEDs

Jumper Wires

A jumper is a short piece of insulated wire that has had about one-fourth of an inch of insulation stripped off the ends so you can insert it into a breadboard and make an electrical connection. Stripping off too much can cause unintended electrical connections, so take care not to expose too much of the conductor. Premade jumpers can be purchased, or you can create your own by using a wire stripper. The size of wire that works best for most breadboards is 22-gauge.

Start with a piece of 22-gauge wire. To strip the wire, insert it into the stripper where the marking shows 22, and then squeeze the handles together and pull down and away from you. Squeezing the handles together cuts the insulation, which should then slide off easily. If it doesn't, try twisting the stripper from side to side to cut the insulation. Take care not to compress or nick the wires as you strip them because that would compromise the conductivity of the wire.

Wearing safety glasses while cutting and stripping wires is a good idea because occasionally the cut wires will fly in an unintended direction, and avoiding eye injury is important (Figure 2.20)!

TIP   When stripping insulation from very short wires, using needle-nose pliers to hold the opposite end will make the job easier.

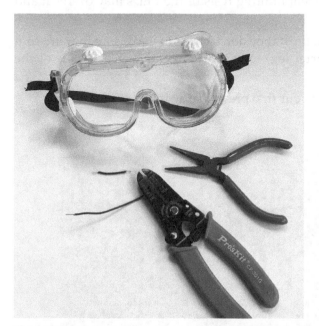

Figure 2.20: Jumpers, wire stripper, needle-nose pliers, and safety glasses

WARNING   Safety Tip! Only use a tool designed to strip wires to avoid injury, and wear safety glasses when cutting or stripping wires!

Building the Circuit

Figure 2.21 shows what the simple breadboard circuit should look like when completed. Follow these steps to build this simple LED circuit:

1. Connect three LEDs as shown in Figure 2.21 by placing the negative (shorter lead) of one to the positive (longer lead) of the next.

2. Jumper the ends of the LED row to the power rails by inserting one end of a jumper in the same row as the last LED lead, one on the positive end of the row of LEDs and another on the negative end of the row of LEDs.

Ensure that the positive end of the LED row is connected to the positive rail and the negative end of the LED row is connected to the negative rail.

3. Briefly connect a 9V battery to the power rails, making sure to put positive in the red rail and negative in the blue or black rail. If you're using a battery snap like the one shown, it's simple. Place red to red and black to black or blue. I strongly suggest you follow this color convention with all your electronic circuits. Note: Connect the power for only a few seconds because without a current limiting resistor the LEDs may overheat and become damaged.

4. The LEDs should light up. If they don't, check to make sure they are connected negative (shorter leg) toward the negative (black) side of the circuit.

5. Ta-da! You just created your first breadboard circuit.

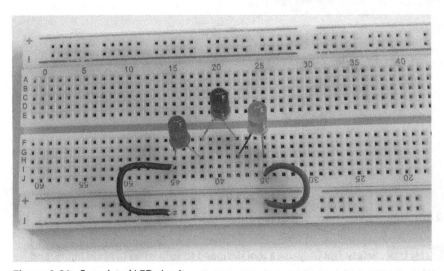

Figure 2.21: Completed LED circuit

Mentally trace the path that the electricity takes from the negative side of the battery, through the rail, and through the jumper to the first LED. Remember, all the holes in that terminal row are connected, so the electricity will flow from the jumper wire across the bottom of the metal clip and into the LED and then through the LED and on to the next one. If you were to put both legs of a component in the same terminal row, the component would short out, would not work, and could possibly become damaged. In Figure 2.22 the red LED is shorted out. Continue tracing the flow of electricity back to the positive side of the battery. Tracing the flow of electricity through the circuit will help you later in troubleshooting circuits that are more complex.

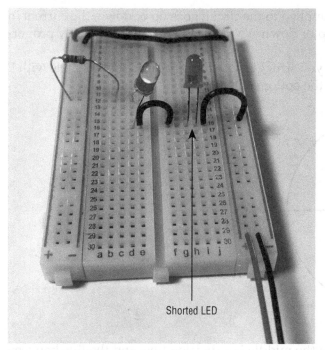

Figure 2.22: LED shorted out

The Basic Circuit

So far, you've learned that a circuit must form a complete path from negative to positive to work and that it must have a source, a load, and conductors. There are some other requirements for circuits to work, which are actually "laws" circuits adhere to, and we must be aware of, when planning circuits. The first is *Ohm's law*.

Ohm's Law

Ohm's law describes the relationship that must exist between voltage, current, and resistance in a circuit for the circuit to work. Knowing Ohm's law is important because using it will help you to choose the right components for a particular circuit. If there is too much voltage or current on a circuit, it could possibly damage the circuit components, or even cause a fire. Knowing Ohm's law and using it to calculate circuit values will help you avoid such dangers.

Ohm's law is simply $E = IR$, where current times resistance equals the voltage applied to the circuit. It is important to note that resistance doesn't typically change (unless the components are exchanged or they're the type that allow resistance to be adjusted, such as potentiometers or light-dependent resistors

{LDRs}), so if the voltage applied to the circuit goes up or down, the current in the circuit will also go up or down. Voltage and current are directly proportional to each other.

A simple trick to use if you struggle with formulas is to draw a circle with E at the top and I and R on the bottom (Figure 2.23).

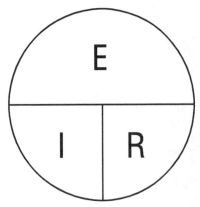

Figure 2.23: Ohm's law

Assume the line across the middle is a division line, and the line between I and R is a multiplication symbol. Cover the characteristic you are trying to determine, for example, current. When you cover the letter I, E over R is left, so E divided by R will yield the current value. If solving for voltage (E), one is left with I times R. If solving for resistance (R), one is left with E divided by I.

Resistor Values and Voltage Dividers

Unless a material is a superconductor or cooled to incredibly low values, it will have at least a little bit of resistance. Conductors have a very low resistance to electrical flow and insulators have a high resistance, but sometimes you need to add a specific resistance to a circuit for it to work. This should be evident now that you know about Ohm's law, and it's done using a component called a *resistor*. Numerous types of resistors exist, but the ones used most often are carbon film resistors (Figure 2.24).

The colors on these resistors aren't just to make them pretty—they have specific meanings. Most have three colors on one end close together and another color, usually gold or silver, on the other end. First, look at the end with three colors. The first two color bands are just a number. The resistor color code chart in Figure 2.25 will help you figure out this number. If the resistor's first two bands are yellow and violet, the resistor's value starts with 47 because the value of yellow is four and the value of violet is seven. However, the third band must also be considered. The third band is a multiplier. Think of it as the number of zeros that are added to the first two numbers. For example, if the third band was black because the value of the color black is zero, no zeros would be added, and the value would simply be 47 ohms. If, however, the third band was orange, because the value of orange is three, three zeros must be added, making the value of the resistor 47,000 ohms.

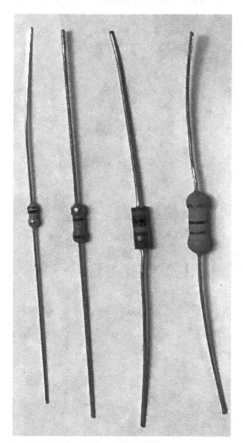

Figure 2.24: Carbon film resistors

The band by itself on the other end of the resistor is a tolerance band. Typically, these bands are silver or gold, but other colors can be used as well. Tolerance indicates the range that the resistor can be tested to and still be considered good. The bigger the value of the resistor, the more important the tolerance is. For example, on a resistor with a value of 470 ohms and a 10% tolerance, 10% of 470, or 47 ohms, would be added to and subtracted from 470 to arrive at the resistor's good range. Meaning, this resistor could test anywhere between 423 ohms and 517 ohms and still be considered good. The total variance possible is 94 ohms. On most circuits, a few ohms doesn't make a lot of difference. If the resistor was much larger, for example, 47,000,000 ohms, then the good range would be 42,300,000 to 51,700,000. The difference in that range is 9,400,000 ohms. That difference could be a matter of life and death, so generally the larger the resistor value, the lower the tolerance will be and the more precise the resistor will be. Silver indicates a 10% tolerance, gold indicates a 5% tolerance, and brown indicates a 1% tolerance. No band usually means the resistor has a 20% tolerance.

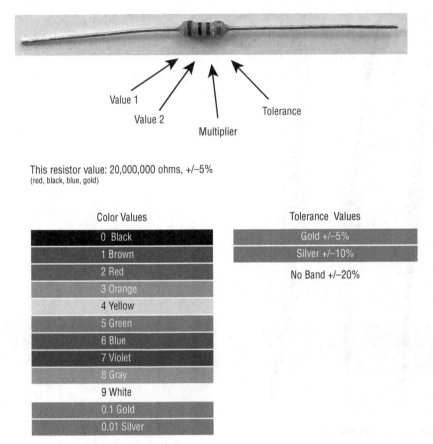

This resistor value: 20,000,000 ohms, +/–5%
(red, black, blue, gold)

Figure 2.25: Resistor color code

Some resistors have other bands as well. On resistors that are extremely precise, you may find four bands instead of three. There may also be bands that indicate the temperature that the resistor is able to tolerate. Resistors also have a total power rating, but that will be discussed in Chapter 6, "Feel the Power."

Resistors can also be used to divide a voltage. Let's say I have the circuit you see in Figure 2.26 with an applied voltage (V_{IN}) of 12V and two resistors in series with each other, going to ground.

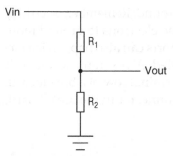

Figure 2.26: Voltage divider

Assume you have a current of 0.001A. Ohm's law tells us that your resistance on the circuit would be 12,000 ohms. If the first resistor, R_1, was 10,000 ohms, and the second resistor was 2,000 ohms, the voltage available at point V_{out} would be 2 volts because the other 10 volts would be changed to heat through the resistor R_1.

Most of the circuits in this book are direct current (DC) circuits, but if an alternating current (AC) such as a sound wave needed to be captured, a capacitor could be used to replace R_2, making the circuit a low-pass RC filter. (R is for resistance; C is for capacitance.) This circuit type would be used to enable low frequencies to pass like those output by a woofer speaker, while blocking (absorbing) the higher frequencies. The values of the resistor and capacitor determine the frequency that will be allowed to pass. That calculation, however, will be saved for another book.

Opens and Shorts

In troubleshooting circuits, the words *open* and *short* have specific meanings. For open, think of an open switch. If the switch is open, no electricity can pass because the conductor (the switch) is open. Therefore, open means there is a break somewhere in the circuit—there is a point at which the resistance becomes infinite and the electricity cannot pass. This could happen, for example, if a resistor's power value is exceeded and the resistor "burns out." Usually, such resistors can be easily spotted because they literally look burned. The resistor can no longer enable electricity to pass and the circuit stops working. It can happen when other devices are broken, too.

The word *short* should conjure up images of wires that have come loose or perhaps dust bunnies. Just as a stray wire can make an inadvertent connection, the dust that often collects in devices such as a computer chassis can have conductive particles in it, which can enable electricity to travel in a path that it was not intended to, bypassing components or parts of a circuit. We call this a *short circuit* because the electricity took a shorter path to ground. Remember, electricity is lazy. If the unintended but available path gets the electrons to ground more quickly, that's the path that they're going to take. Shorts can also happen if a circuit has been wired incorrectly and should be avoided. An example would be if both leads of an LED were inserted into the same terminal row of a breadboard. A short would exist across the LED via the metal connector in the breadboard, and the LED would not light. (See Figure 2.22.)

Circuit Protection Devices

Some devices are placed into circuits with the intention that they will break under certain conditions. Fuses and circuit breakers are two such devices (Figure 2.27).

Figure 2.27: Fuses and a circuit breaker

Fuses

A fuse is a relatively simple device. It is often simply a piece of wire encased in a glass tube. The *fuse* is designed to melt and break apart when a certain current is exceeded. The purpose of the fuse is to protect other devices on the circuit from being damaged, or to mitigate injury to people using the circuit when they inadvertently come into contact with the circuit and excessive current is flowing *through them*. The fuse breaks, current no longer flows, and damage is mitigated. A fuse can prevent explosion or fire in a circuit or a tool used to measure electricity. If a fuse has blown, you may want to inspect other devices on the circuit for damage before putting the circuit into use again.

Fuses, like most devices, can handle short bursts of electrical power that are higher than their rating. Both sustained and pulsed characteristics need to be taken into consideration when choosing components. For example, a fuse rated for 1A may be able to withstand a current of 1.5A or 2A for a few milliseconds or minutes, depending on the fuse. Some fuses are designed to be fast acting, meaning they will break the circuit quickly, while others are designed to have a longer lag time. In some applications, a longer lag time is desirable. Ambient temperature also affects the characteristics of a fuse. Information about how quickly a fuse will break should be taken into consideration when choosing a fuse for a project and can be found on the fuse's datasheet. Datasheets are generally available on the manufacturer's or vendor's website, and contain specifications and limitations for the component.

A multimeter is a device for measuring characteristics of electricity such as current and voltage. Multimeters have fuses within them to protect electricians, electronic enthusiasts, and the meter itself from being damaged by too much current. (If a meter isn't working properly, check the fuse. Replacing it might solve the problem.) Many fuses can be found in automobile circuitry. The somewhat square fuses shown in Figure 2.27 are automotive fuses. Many devices will have more than one fuse in them. How to test a fuse is explained in Chapter 5, "Dim the Lights."

Circuit Breakers

A circuit breaker is similar to a fuse in its purpose but is much more complicated in its construction. The larger device shown in Figure 2.27 is a circuit breaker, which can be found in the electrical panel of a building. If a building is very old, it may use fuses like the round yellow one in Figure 2.27 instead of a circuit breaker. *Circuit breakers* are designed to "trip off" when a certain current is exceeded. One method to achieve this is by using a bimetallic strip. Metals will expand at a certain ratio when heat is applied. Current flowing through a resistance will always produce some heat, and even good conductors like metals provide some resistance. A bimetallic strip will have two metals that expand at different rates, causing the strip to curl in one direction or the other. The more heat added, the more the metal expands. When it reaches a certain point, a switch inside the circuit breaker is tripped off, which stops the current from flowing. The advantage of circuit breakers is that they can be easily reset and used more than once, unlike a fuse, which must be replaced each time the current is exceeded.

Bigger Is Not Better

When a fuse must be replaced, it's important to understand that bigger isn't better. If a circuit has a 1-amp fuse and it blows, it should *not* be replaced with a 2-amp fuse. One might think that a bigger fuse would be fine, but if a circuit designed for only 1 amp of current was allowed to reach 2 amps, it would likely damage components on the circuit and could even cause a fire due to overheating. Again, knowing the fuse's characteristics related to pulsed versus sustained current, ambient temperature and lag time are important and can be found on the fuse's datasheet.

Similarly, if the normal current on a circuit was expected to be 1 amp and a 1-amp fuse was a part of the circuit, the fuse might repeatedly burn out and need constant replacing. A fuse should have a value slightly higher than what the circuit is expected to use, or an appropriate lag time, but low enough value so that the components won't be damaged before the fuse breaks the circuit. Fuse values are usually printed on the fuse casing and note the maximum voltage as well as current (Figure 2.28).

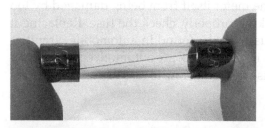

Figure 2.28: Imprint on a fuse casing

WARNING   Safety Tip! Always replace a fuse with one rated for the same voltage and current.

Symbols and Diagrams

"Human nature is to need a map. If you're brave enough to draw one, people will follow."

Seth Godin

If you're going to California from New York and you head north, you're going to run into Canada. Because Canada isn't your destination, you likely need a map. Whether your map is a paper one like your grandfather had, a global positioning system (GPS) device like your mother had, or the Maps app on your phone, you need to know what direction to go and where you can get gas or sleep along the way. You need a map. This chapter looks at different types of maps used in electronics and some of the common symbols you need to know right away. You'll learn more of them as you go along.

Types of Diagrams

Several different types of diagrams are used in electronics. *Wiring diagrams* show how devices are physically connected. When you see a drawing of a breadboard and wires going here and there among components, that's a wiring diagram. Boxes may be used to represent a component, or the wiring diagram may be combined with a pictorial diagram.

Pictorial diagrams, as the name implies, may have actual pictures of parts and are used for identifying those parts and where they are physically placed.

Block diagrams show the big picture. Complex projects may have a block diagram that simply has a box for each major section of a project to show how the sections are connected. Each section will have its own detailed diagram called a schematic.

Schematic diagrams use symbols to represent components, and while the symbols might not look at all like the actual component, they will show how the components are connected electrically. Schematics are the "maps" most often used in electronics. Figure 3.1 shows a block diagram and an expanded section showing the schematic for that section of the block diagram.

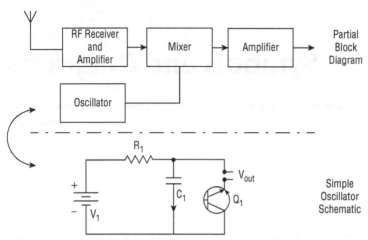

Figure 3.1: Block diagram and schematic

Look at the symbol on the bottom left of the schematic in Figure 3.1. This is a battery symbol, and the longer line indicates the positive side of the battery. Putting the positive side of a circuit on the top of the drawing is a common practice. If you orient your breadboard so the positive rail is at the top, it makes converting from a schematic to a real circuit much easier. Schematics are also commonly set up to be read from left to right, and because everything starts with power to the circuit, that is usually where you will find the *source* of electrons whether it is a battery or *mains power* (an AC wall outlet).

Notice that the battery symbol is labeled V_1. V for voltage, and 1 means it's the first voltage source in the schematic. A battery might also be labeled B_T, or V_{IN} (voltage in), or simply have the voltage shown such as 12V. Typically, each component in a schematic will have a letter/number combination so it can be identified. When explaining a circuit to someone else or reading reference material, it's much easier to say C_1 for the first capacitor instead of "the capacitor that is connected between the first resistor and ground." Many components start with the same letter, such as switch and source. S is generally used for a switch, so a different letter is used for source. Table 3.1 shows commonly used letter designators.

Table 3.1: Commonly Used Electronics Designators

COMPONENT	SCHEMATIC DESIGNATOR
Diode	D
Light-emitting diode (LED)	D (or LED)
Inductor	L
Capacitor (nonpolarized)	C
Electrolytic capacitor	C
Basic switch	S (or SW)
Resistor	R
Variable resistor	VR
Transistor—BJT	Q
Transistor—MOSFET	Q
Battery (DC source)	BT (or B or V)
Mains power (AC source)	AC (or V)
Integrated circuit chip	IC

Notice that L is not used for LED. L is the symbol for an inductor, which in its simplest form is a coil of wire that holds a magnetic field. T is not used for transistor because it is used for transformers.

When electronics were first being invented, the scientific community thought that energy moved from positive to negative. That idea is called the *conventional current theory*. We now know more about atomic structure and know that current flow is from negative to positive as electrons rush to fill the holes left in atoms by missing electrons. This currently accepted theory of current flow is called the *electron flow theory*. Often circuits are still drawn following conventional current theory. Keep this in mind as you encounter circuits drawn by other people.

Schematic Symbols

Knowing commonly used symbols is imperative when trying to understand what a schematic is illustrating. A set of standard symbols has been defined by the American Society of Mechanical Engineers (Y14.44) , and another by ANSI/IEEE (315a). ANSI is the American National Standards Institute, and IEEE (pronounced eye-triple-e) is the Institute of Electrical and Electronics Engineers. Both of these groups are well recognized and their standards accepted.

NOTE   More information about each of these groups is available on their respective websites, and you can even sign up for membership. Find these groups at `ASME.org`, `ANSI.org` and `IEEE.org`, respectively.

Most often the symbols encountered will at least be similar to those standards, although new components are being developed daily. Note, a difference exists between symbols commonly used in the United States and symbols used in Europe. Some symbols have changed over time so you may encounter older versions and newer versions (e.g., conductors). The symbol for a conductor (usually a wire or breadboard connection) is simply a line. In complicated schematics that could be a problem because it may be impossible to draw a conductor without it crossing over another conductor. How would you know whether they are connected? See Figure 3.2.

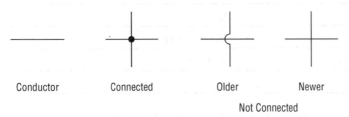

Conductor   Connected   Older   Newer

Not Connected

Figure 3.2: Conductor symbols

The dot where two or more conductors are joined is called a *junction*. Junctions may also appear where two or more components are connected. Consider that each terminal row of a breadboard (numbered row with usually five or six holes) is actually an electrical junction. When you see the dot, think "terminal row."

Figure 3.3 shows the most commonly used schematic symbols. Some of them haven't been discussed yet. No worries! Each of them will be discussed at length in the coming chapters, and more symbols will be added. There are far too many symbols to include all of them here.

A symbol for an integrated circuit chip is not included because the shape changes depending on the type of IC that is being used. For more details about integrated circuit chips, check out Chapter 13, "Integrated Circuits and Digital Logic."

Some of the symbols, like the light-emitting diode and photodiode, have two arrows. The arrows indicate light, either coming into the device and affecting it in some way or going out of the device. There are photovoltaic cells that take light and turn it into electricity, photodiodes, photoresistors, and so on. Just remember that when you see the two arrows together, they indicate light.

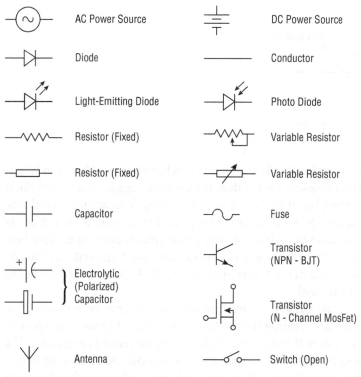

Figure 3.3: Common schematic symbols

So Many Switches!

The symbol for a basic switch was shown in Figure 3.3, but there are numerous different types of switches; interpreting what the symbols all mean can make a dramatic difference in the functioning of the circuit.

First, let's consider the parts of a switch. (See Figure 3.4.) A switch has one or more poles. A pole is a common connection point that stays connected to the circuit regardless of what is happening on the other end. A switch also has one or more throws. A throw is the part of the switch that connects to one or more circuits. Figure 3.4 shows a switch with two throws. The actuator is the part that moves to either complete (make) or disconnect (break) the circuit. An abbreviation such as SPST simply means single pole, single throw. With an SPST switch, only one circuit can be controlled by either turning it on or turning it off. This is a simple toggle switch, like most light switches, and the switch schematic is shown in Figure 3.3. A switch can be SPDT, meaning it can activate (turn on) two circuits but only one at a time, as in Figure 3.4. A switch could also be DPDT, meaning there are two constant connection points and two throws for each, so a total of four circuits are controlled, two at a time.

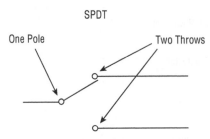

Figure 3.4: Poles and throws

All switches have an actuator. The *actuator* is what changes the switch between the open (off) and closed (on) states. It may be a toggle that is switched from one side to another, but it can be almost anything. The actuator could be a magnetic field that causes a metal bar to be pulled in a specific direction as it is in a relay. It could also be a drop of water that causes current to flow between two points, or sunlight hitting a photosensitive switch. Switches can be attached to the sensors that make Internet of Things (IoT) devices do things, so switches are a rather big deal.

Open (off) switches have broken the electrical connection between two points, and closed (on) switches are creating the electrical connection between two points. A switch's *at-rest* position is the state it is in *before* the actuator is activated. If a circuit is off until you activate the switch, the at-rest position of that switch is normally open, represented by NO. If the circuit is on *until* you activate it, then the switch is normally closed, which is represented by NC.

SAFETY TIP!

Choosing an NO or NC switch may affect more than the function of the circuit; it might make a difference in safety. Never assume a switch is NO or NC until you've tested it.

A switch can be momentary, meaning that it's activated only as long as the button is held or a condition is in place that keeps it activated. When the condition no longer exists, the switch returns to its at-rest position. For example, a circuit whose switch is activated by a water level reaching a certain height may stay on until the water recedes below the activation point. The switch is a momentary NO switch. The rising water activates (closes) the switch, enabling electricity to conduct. The switch continues in that state as long as the water is above a certain point. However, when the water level drops again, the switch returns to its at-rest position, which is open (off).

Figure 3.5 shows just a few of the available switches. A website like digikey.com or mouser.com will show tens of thousands of switches. For example, a query of switches on digikey.com today showed more than 172,000 pushbutton switches, and that's just one of more than a dozen types of switches listed.

Figure 3.5: A sampling of switches

The brass-colored switch in the front in Figure 3.5 is a motion-detecting switch with a small amount of mercury encased in a tube. If the tube is tipped far enough to the right, the mercury will connect two small conductors inside that connect to the leads outside and complete the circuit.

Starting at the next row on the left, Figure 3.5 shows two round switches, one black and one red. These are both SPST pushbutton momentary switches. Next to them are two pushbutton switches that are not momentary (i.e., once pressed, the circuit stays in the activated state until pressed again). Next up are two small slider switches. They both have a position in the middle where everything is off, but sliding the actuator left or right will turn on one of two circuits. Some slider switches don't have this center off position. On the far right is a dual inline package (DIP) switch. The same effect could be achieved with four SPST switches.

In the next row starting at the left, you'll find an SPST toggle switch. The second switch is an SPDT toggle that can be lit up. Next is a tiny DPDT toggle and then a 2/12 rotary switch, which has two poles that control six circuits each. At any given time, two of the six circuits will be closed. Turning the knob opens the current set and closes the next two connectors. Finally, there's a keypad that is simply a dozen push-button switches connected together. Typically, a keypad switch will have a common (shared) anode (positive) or cathode (negative) connection, in other words, 1 pole and 12 throws, each activated by pressing a different number on the keypad.

> **NOTE**   Some switches are designed to fit directly into a breadboard, like the STSP toggle switch in the back row on the left. Notice that it has small, solid leads which are spaced to easily fit into most breadboards. The two switches to its right have leads with holes in them. The holes are for looping hookup wire through and then soldering the wire to the switch's lead. These leads are called solder lugs. The solid leads for breadboards are called breadboard lugs.

Different switch types have different schematic symbols. Figure 3.6 shows the most common of these schematic symbols.

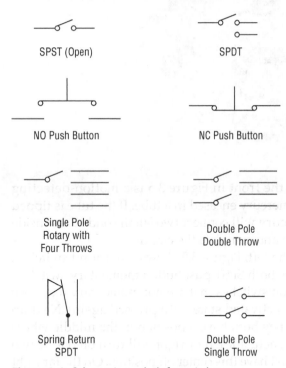

SPST (Open)   SPDT

NO Push Button   NC Push Button

Single Pole   Double Pole
Rotary with   Double Throw
Four Throws

Spring Return   Double Pole
SPDT   Single Throw

Figure 3.6: Schematic symbols for switches

A final consideration for switches is the arcing that occurs when a switch is opened or closed. Arcing happens when the pressure of electrons to connect to the positive side of a circuit overcomes the resistance of the material (usually air) between the two connection points. Electrons jump across the gap, and arcing occurs. Arcing is what happens when you walk across a carpet in winter and reach toward a metal doorknob. You may see a flash of light, hear a cracking sound, and feel pain as the electrons rush toward the doorknob. Switches undergo these same forces, so they must be manufactured in such a way that they can withstand arcing without failing. Switches, and in fact most electronic devices, are rated for the current and voltage that they can safely handle. If a circuit is alternating current and 240V, then you would want to make sure a component is designed for that at a minimum, and generally you would want to go a bit higher just to ensure failure doesn't happen.

Specifications for devices can be found on a *datasheet* for that specific device. Figure 3.7 is an excerpt from the manufacturer's datasheet for a series of toggle switches.

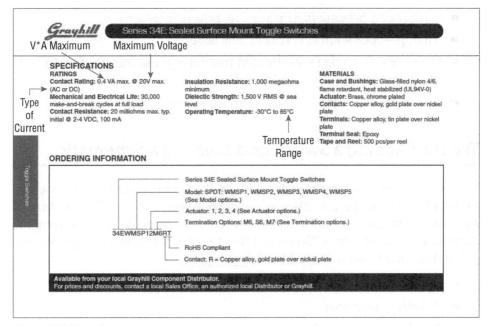

Figure 3.7: Datasheet

Notice the maximum voltage for this device is 20 volts, but it can operate on either AC or DC. The volt amperes (VA) can't exceed 0.4, which means if it were operating at 20 volts, the current would be limited to 0.02 amps. This datasheet even specifies the temperature range under which the switch will operate properly.

To find a component's datasheet, try the manufacturer's website or the website of the supplier from which the item is being purchased. Most datasheets are fairly easy to find because the manufacturer wants to make it easy for you to purchase their products.

Drawing Your Circuit

Hand drawing circuits on paper is fine for a simple circuit, but for something more complex it's usually easier to use a software program designed to create circuits. The advantage of these programs is that they will have the schematic symbols available to choose from a list, and changes are easier to make than erasing and redrawing the circuit. Many free programs are available online, some from suppliers, some from software companies, and others from various enthusiast groups. When using any of these resources, be sure to read and adhere to the licensing, user, and privacy agreements. Some really are free, but others assume ownership of anything you save on their website, so choose carefully. Here are a few resources:

- `Digikey.com`, Scheme-it: `digikey.com/schemeit/project/`
- Fritzing—open source: `fritzing.org/home/`
- `Autodesk.com`, Eagle Free—limited free general use but free for students and educators: `autodesk.com/products/eagle/overview`
- `Schematics.com`—free to use: `schematics.com/`

Try This: Adding a Switch and Creating a Schematic

Figure 3.8 shows you what you'll need to build a simple circuit with a switch. Most LEDs can only handle 20–50mA of current before they will be destroyed, so whenever you use an LED in a circuit, you'll need to use a resistor to limit the current. This circuit uses a 330-ohm, 1/4-watt resistor (orange, orange, brown).

For this circuit, you will need the following:

- Breadboard
- 9V battery with snap
- 330-ohm 1/4-watt resistor
- LED
- SPST switch
- Jumper wires

Adding the switch is a simple process. First, look at the wiring diagram in Figure 3.9. Instructions for completing the circuit follow.

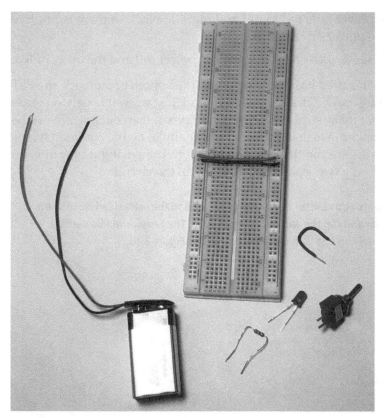

Figure 3.8: What you need

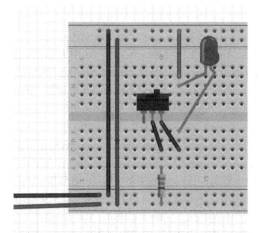

Figure 3.9: Wiring diagram of simple circuit with switch

1. Jumper the power rails red to red and black to black so there is positive and negative power on each side.

2. Insert the resistor with one end in the red power rail and the other in 10a.

The physical switch used has just two pins on the bottom because it's an SPST switch, but an SPDT or DPDT switch would work just as well if only one side of it were used. Just make sure to use the center pin and an outside pin on the same side. The center pin is usually wired internally to be the common (pole), and the outside pins are the throws. The image in the wiring diagram is an SPDT switch, and only two leads are connected in the circuit.

TIP   If you're putting a resistor in a narrow space, bend the metal lead close to the body of the resistor and clip the ends so they are even. The resistor will be vertical instead of horizontal and can fit in a smaller space. See Figure 3.10.

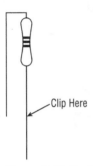

Clip Here

Figure 3.10: Vertical resistor

3. Insert the switch so that the middle lead is in 10c. The outside lead (or other lead if only two) should then be in 12c. Depending on the switch you are using, you may be able to insert the switch directly into the board like the one shown in Figure 3.11, or you may need to solder leads onto the switch like the black momentary button shown in Figure 3.5.

4. Insert the LED so that the longer, positive lead (anode) is in 12e, the same row that the outer lead of the switch is in. Put the other lead (negative, cathode) in 12h. On the breadboard in Figure 3.11, the numbered rows are opposite on opposite sides of the board, so it looks like the lead is in 48g. Note, regardless of how the breadboard being used is numbered, the dip in the middle is a break in the metal connections beneath the board, so abcde of the same numbered row is one junction and fghij is another junction.

5. Put a jumper in 12i (or 48i depending on the breadboard being used), and insert the other end into the negative power rail.

6. Plug the power from the 9V battery into either set of power rails, being sure to match red to red and black to black or blue.

Flip the switch and. . .ta-da! Let there be light! In the unfortunate event that there isn't light, check that all your connections are actually connected. It's easy to bump one component or wire while putting in another and pop the first component out of the hole it should be in. Try flipping the switch in the opposite direction. Follow the electricity around the circuit from negative to positive, checking each connection along the way. Make sure the LED is oriented properly and that the LED isn't blown. For this purpose, consider keeping a 3V battery around (like the flat ones often found in watches) to test LEDs with. If the LED works with the battery, the problem is not the LED, but it may be that the leads are reversed. You may find a pin isn't in the right row. It's okay. It happens and can be fixed.

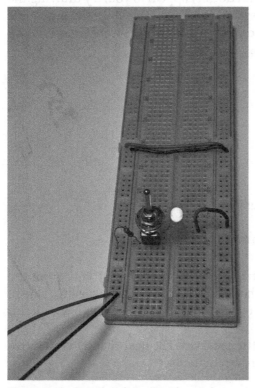

Figure 3.11: Completed circuit with switch

TIP   Keep a 3V watch battery handy for testing LEDs. Touch the positive side of the battery to the longer lead of the LED and the negative side of the battery to the shorter LED lead. If it lights, the LED is good. A 9V battery won't work for this because there's too much voltage and the LED will be destroyed.

Finally, it's time to make a schematic of this circuit, but keep in mind that usually the schematic is created first, and the circuit after. Remember that a schematic shows how components are electrically connected and might not look at all like the actual circuit does. Schematics use symbols, not pictures or drawings. Figure 3.12 shows a possible schematic. Yours could look different but still be correct as long as the connections are correct. Choose a program that appeals to you from the ones listed earlier in this chapter, or find another that you like working with and use it to create a schematic for this circuit. If you can find one that will simulate your circuit running so you can tell if it will work or not, it's a great tool to have. If you've never created a schematic before, it's good to practice with more simple circuits like this one and work your way up to harder circuits. If using a program doesn't appeal to you, try drawing the circuit with paper and pencil but keep an eraser close by. Both the schematic shown in Figure 3.12 and the wiring diagram in Figure 3.9 were created using the circuit diagraming tool from `Fritzing.org`.

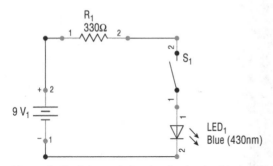

Figure 3.12: Schematic for simple circuit with switch

When you're done, remember to trace the path of the electricity from the negative side of the source, through the circuit, and back to the positive side of the source. Thinking about how the electricity is flowing through the circuit will help you find any errors in it.

While this is exciting, we're just getting started. The next chapter is an introduction to the amazing Arduino board, so read on.

Introduction to the Arduino Uno

"You are responsible for what you say and do. You are not responsible for whether or not people freak out about it."

—Jen Sincero

When I was a little girl, people would give me doll s to play with. My brother's toy trucks with their motor sounds and lights were much more interesting, so inevitably the dolls were set aside, and I took apart my brother's trucks. I wanted to know how everything worked. Back then, little girls with curly blonde hair and blue eyes were expected by society to be sweet and quiet and follow the rules . . . . I did not comply, which might explain my fondness for Arduino and its mission.

The folks at Arduino don't follow conventional rules either. They made their hardware open source, so virtually anyone can build, use, and modify the Arduino system. Their "ecosystem" of hardware and software, along with a community of innovators and enthusiasts, makes learning electronics and being creative easy and fun for nonprogrammers and nonelectronics engineers—like you, perhaps? So, what is Arduino?

What Is Arduino?

Arduino, it turns out, can refer to a lot of things. Arduino is a company (`Arduino.cc`) that has circuit boards and related materials for sale, but when people speak of Arduino, they're more often referring to the myriad of boards and design software enabling a person to create a program and upload it to a microcontroller on an Arduino board. The hardware and software collectively are the *Arduino environment*.

The Arduino environment is widely used by enthusiasts to create interactive tools and playthings. It's also used by companies to create prototype devices for Internet of Things (IoT) uses. The IoT market has barely started growing, so you're in a great place right now to take advantage of that IoT market growth and show off your creative genius. Learn to prototype on an Arduino board, allow yourself to bend the rules, and get creative. Who knows where this could lead!

The board we're using in this book is the Arduino Uno. When you connect an Arduino board to an electronic circuit, the board can accept electrical input from all sorts of sensors. These electronic circuits convert inputs from sensors, such as water level or temperature, to an electrical voltage that changes as the input changes. This voltage is then handed over to the Arduino board's input pins and processed by the microprocessor based on a program that the user created and uploaded to the microprocessor.

Beyond the basic hardware and software, numerous *Arduino shields* are available that extend the ability of an Arduino board. The shields save time because they provide common features like Wi-Fi, a global positioning system (GPS), or a liquid crystal display (LCD). Hundreds (maybe thousands) of shields are out there, but it's fun to design your own. Shields also often include example *sketches* or a *library*.

A sketch is the Arduino term for a program. A program consists of lines of code (instructions) that are written in the Arduino *integrated development environment* (IDE) to be uploaded to the microcontroller.

A library consists of lines of C or C++ programming that add functionality to sketches and can be imported into any sketch. A library can be created by anyone, even you!

With all that in mind, when someone says they have an Arduino, it could be one of several different circuit boards with a microcontroller that can do, or interact with, almost anything. It could even be a board that they have built with the design provided on the Arduino.cc website. Before you decide to build your own Arduino board, consider that building one is not a beginner project. It's better to learn about electricity, electronics, and the Arduino IDE first.

The Arduino Board

As previously mentioned, the Arduino board used the most in this book is the Arduino Uno (Figure 4.1). It's small enough to fit in your hand, but it's still a powerful board and likely the one that most people use to begin learning the Arduino environment.

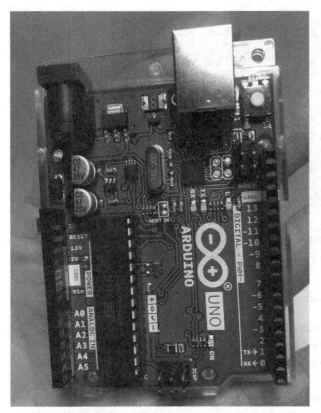

Figure 4.1: Arduino Uno board

The following is a quick overview of the physiology of the board. While it may seem like a lot to remember now, remembering the parts of the board and what they do will become much easier as you create circuits.

First, check out the ATmega328 microcontroller (see Figure 4.2). This chip executes the commands from the Arduino sketches. It has 32kB of memory (32 kilobytes, or 32,000 bytes); however, only 31.5kB is available for programs. The other 0.5kB is used for the bootloader, which is embedded programming that starts the system working.

The commands for the microcontroller are uploaded from your computer through the USB port. This USB port can also be used to power the Arduino. USB ports provide 5VDC and 500mA, which is 0.5A. The operating voltage of the Arduino board is 5VDC.

   Microprocessor

Figure 4.2: ATmega328 microcontroller

> **TIP** DC stands for "direct current," like what is supplied by a battery. With DC, the electrons flow in only one direction. AC stands for "alternating current" and is typically supplied by a wall outlet or a generator. Electrons in alternating current flow first in one direction and then the other. In the United States, a wall outlet supplies approximately 120 volts of electricity, so a transformer must be used to step down the voltage to the proper voltage for the device being connected. For an Arduino Uno, the recommended input voltage is 7VDC to 12VDC when using the V_{in} pin. Transformers are explained in Chapter 16, "Transformers and Power Distribution."

The Arduino Uno board has a fuse to protect the computer's motherboard against too much current going through the USB port and back into the computer. If the current exceeds 500mA, the fuse will open (turn off) the connection, preventing the excess current from reaching the computer. Unlike many fuses, this fuse is resettable and will close (turn on) the circuit when the current returns to the acceptable range.

A green LED on the right side of the Arduino Uno board indicates that the board is receiving power. On the left of the board, you will find transmit (TX) and receive (RX) LEDs that indicate when data is being received into or sent out from the microprocessor (Figure 4.3). These LEDs flash rapidly during data transmission, such as when a sketch is being uploaded to the microprocessor.

Other power options are a power jack and power pins. The power jack accepts a 2.1mm center positive plug and connects to a wall outlet through a transformer like the ones used for most any other electronic device. It can also connect to a 9-volt battery with a proper adapter (see Figure 4.4).

USB
Port

Power
Indicator
LED

Transmit
and
Receive
LEDs

Power
Jack

V_{in} Pin

Figure 4.3: TX and RX lights and USB port

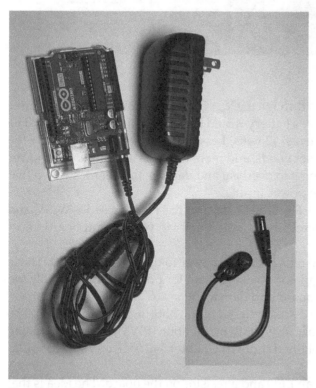

Figure 4.4: Adapter for 9V battery and a transformer

If you purchase a power transformer to go from a wall outlet to your Arduino board, make sure that it meets the specifications of the board. You might even have an old one from another device, such as a cell phone, taking up space in a drawer or closet. A cell phone transformer provides 5VDC and can be connected through the USB port. If using another transformer, it should connect via the power jack. Make sure you test the transformer's output voltage first. (Measuring electricity is covered in the next chapter.)

If you do use the 2.1mm jack to supply power to your board, the V_{in} pin can be used to access that power (Figure 4.5).

Figure 4.5: Power jack and USB port

Power can also be supplied through the V_{in} pin and the ground pin in the Power section of the Arduino, where a battery would be connected (see Figure 4.6). The recommended input voltage is 7–12 volts. Less than that might not provide the 5 volts that the microprocessor needs, and more than 12 volts might overpower the voltage regulator, causing it to overheat and damage the board, so keep that in mind when planning the power inputs.

The section labeled POWER on the board also has pins labeled 3.3V, 5V, and ground. These pins are intended to supply power to your circuit board, most likely a breadboard.

Don't use the 3.3V and 5V pins to supply power to the Arduino board. Those pins bypass the voltage regulator, so using them for power in could enable too much power to pass to the board and damage it.

Across the top of the board are 13 digital input/output pins (Figure 4.7). The pins with the ~ are capable of pulse width modulation (PWM) output for analog devices. (PWM will be explained more in Chapter 14, "Pulse Width Modulation" of this book.)

Pin 13 is associated with an LED and a resistor on the board. The idea is that pin 13 can be used for troubleshooting without involving a circuit board. When the power on the pin is high, the light is on. If power on the pin is low or off, the LED is not lit.

Figure 4.6: Power connections

Next to pin 13 is a ground pin. The ground pin should always be connected first. Across the bottom of the board are six analog input pins. These analog pins can also be configured for use as digital input pins (Figure 4.7).

Figure 4.7: Digital and analog pins, LED 13, and GND

Analog vs. Digital

If the changing voltage of a circuit were plotted over time, the resulting pattern can be described as either an analog or a digital signal. *Digital* signals are analogous to the binary signaling used in computer systems. Either an individual wire is charged (binary 1), or it is not charged (binary 0)—on or off, no in-between. Assume, for example, that a circuit has an output causing a light to flash steadily, with the time on and time off being equal. The resulting pattern is described as a square wave (Figure 4.8). This type of output can also be described as two digital logic states known as *logic high* and *logic low*. Logic high equates to true, binary 1, or on, and occurs when the voltage is at or near its maximum. In the case of the Arduino Uno, logic high is 5 volts. Logic low equates to false, binary 0, or off, and obviously occurs when voltage is at or near zero.

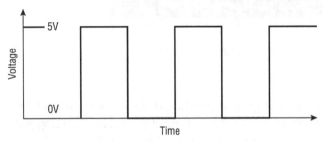

Figure 4.8: A square wave

An analog signal, by contrast, includes the full range of values from 0 to maximum and *everything* in between (Figure 4.9). Imagine the ripples caused by a rock thrown in a pond. That rock produced an analog signal that could be interpreted by a sensor floating on the water. If the height of the wave were converted by a transducer into voltages, then a sine wave would result, although the sine wave would be decreasing in intensity as the wave subsided. The peak of the wave would produce a maximum voltage, the trough of the wave would produce the minimum voltage, and the up and down of the ripples would equate to varying voltages as the wave moved past.

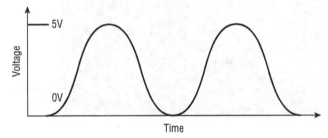

Figure 4.9: An analog wave

This is the essence of IoT. Sensors pick up a change in something and output that information as a change in voltage or current, which is manipulated by some sort of logic device and turned into data. What the sensors receive as input can be in digital or analog format, and the output can be, too, although for the Arduino's microprocessor to manipulate the signal, it needs to be in a digital format.

How does an analog signal become digital? First it's the job of the transducer to convert what the sensor picks up (such as the change in height of the wave) to a voltage. The voltage becomes input on an analog pin of the Arduino, and then an analog-to-digital converter (ADC) chip takes this varying range of analog voltages and changes it into digital data that the microprocessor can use.

A typical ADC like the one on the Arduino Uno might have 10 available binary bits to represent the value, which would give it a range of 0 to 1023. If there were an analog input signal with voltages ranging from 0 to 5 volts, then 0 volts would correspond to digital 0, and 5 volts would correspond to 1023. The numbers from 0 to 1023 can be interpreted and manipulated in the software of the microprocessor. Exactly what it does with those numbers is determined by the person who wrote the code.

TIP   Binary is a base 2 number system. Because the only characters to work with are 0 or 1 (off or on), each placeholder in the binary system represents two possible states. If there are two bits to represent data, the options would be 2^2, (2 x 2), or four possible values: 00, 01, 10, and 11. With 10 bits available, the possibilities are 2^{10}, or 1,024 possible values. Visit `CliffJumperTek.com/binaryisfun` for more information on binary values.

The process in reverse can be used to create an analog output. The device at work in that process is a digital-to-analog converter (DAC). An example of this might be a program for writing music. The writer enters the note that they want played, that note corresponds to a sound frequency (analog wave), and a microprocessor could be programmed to output those analog voltage values to a speaker circuit.

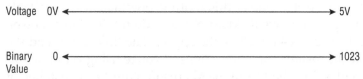

Figure 4.10: Binary values across 5 volts

Finally, the Arduino has a reset button (Figure 4.11). This button can be pressed to reset the microcontroller, which clears out the previously used program before uploading a new one. Another way to accomplish the same objective is to use a software reset. On a Windows PC, this happens when the Upload button is clicked on the IDE toolbar, so each time software is uploaded, the microcontroller is reset. On macOS and Linux machines, it happens whenever the IDE software makes a connection to the microcontroller. It is possible to disable this automatic reset, but for most purposes it works well to leave it as it is.

Figure 4.11: Arduino reset button

The Arduino IDE

The true power of the Arduino lies in the ability to program what happens to the data it receives. The first circuit most people program is a simple flashing LED, but before that, the Arduino IDE must be installed on a computer. The IDE is where we tell the Arduino's microcontroller what we want it to do.

The IDE can be downloaded from `Arduino.cc/en/Main/Software` or via the links at `CliffJumperTek.com/Arduino/`. If you prefer to work online, you can use the Arduino Web Editor. At the time of this writing, downloads are available for Windows XP through Windows 10, Mac OS X 10.8 Mountain Lion or newer, and Linux/Linux ARM 32 or 64 bit. Click the appropriate link, download the software, and follow the installation instructions. If you're using Windows 8.1 or 10, the software will be installed from the Microsoft Store. Once installed, it will appear in your Start menu with the familiar Arduino logo.

When the program is launched, the default sketch screen appears. As mentioned earlier, a sketch is a program for telling the microcontroller what you want it to do. The Arduino IDE uses the familiar Windows setup. Across the top is a menu bar; beneath that is the taskbar. The large area in the middle is for writing the code. As with most text editors, copy, paste, and delete work here. The bar located below the text editor is the *message area* and the *text console area* where the details about the messages generated when a sketch is compiled are shown.

TIP   If the Arduino IDE isn't installed yet, you might want to do that now so you can work with the actual windows as you're reading.

Figure 4.12 shows what the screen looks like after clicking the check mark on the taskbar. This check mark is also known as the Verify button. Notice that the message area says, "Done compiling," and beneath that information about the program on the screen is shown, including how much program space is still available.

Figure 4.12: Default Arduino IDE screen

The IDE uses a high-level programming code that is easy for us to understand, but not so easy for the microprocessor to understand. When the program is compiled, it is converted to machine language, commonly called *binary*, that the microprocessor can understand.

If an error is made in the syntax of the code, clicking the Verify button will give you more information about any errors that are found, as well as suggestions on how to fix them, such as "; expected before Println". *Syntax* simply means the specific order in which things are done, such as putting a space or an opening or closing bracket where it belongs. It can't fix errors in programming logic, but it can make finding some mistakes easier.

Next is the Upload button, which is simply an arrow pointing right. This button will upload the program to the microcontroller, assuming the computer is connected to the Arduino board.

The middle button, which looks a bit like a piece of paper, is for starting a new sketch. The button pointing up is to open an existing sketch, and the button pointing down will save the sketch that is currently in the text editing window.

The default name for the sketch shows on the tab of the text editor, in this case, sketch_nov29a. To rename your sketch, save it by clicking the Save button (down arrow) and then choose a new name. To rename an existing sketch, click the arrow to the far right of the text editor tab, then click rename. A box opens asking for the new filename. Type the name and click OK.

Multiple tabs can be open at the same time, each its own sketch. Simply click the same arrow to rename a file, and choose New Tab. You'll then be prompted for a filename for that new sketch (Figure 4.13).

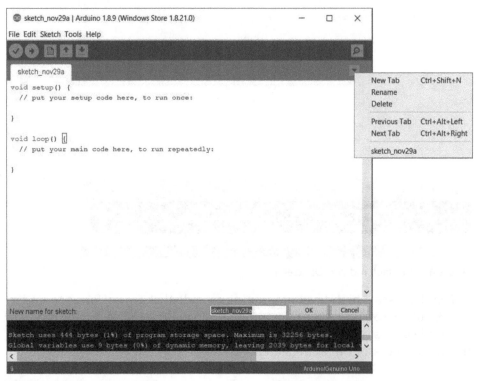

Figure 4.13: Renaming a sketch

TIP   In the IDE, click File ⇨ Preferences. On the Settings tab, you can change the font size, set the file location, and display line numbers to make your coding work easier to follow.

Try This: Creating a Simple Arduino-Controlled Circuit

The best way to learn is by doing, so it's time to work with a first sketch. For this project, you'll need the following materials (Figure 4.14):

- Arduino Uno
- Computer with the Arduino IDE installed
- USB cable
- Breadboard
- LED
- 330-ohm resistor (orange, orange, brown)
- Jumper wires: 1 red and 1 black

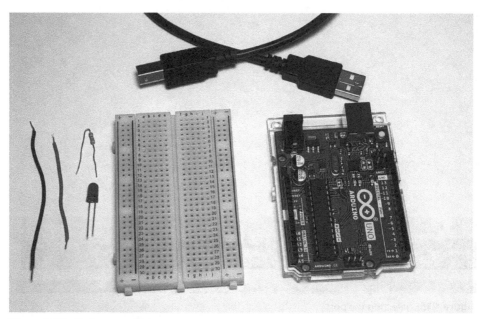

Figure 4.14: Project materials

1. Open the Arduino IDE.
2. Plug your Arduino into your PC via the USB cable.

3. On the menu bar, click Tools, scroll down to Board, and make sure the correct board is selected, in this case Arduino/Genuino Uno.

4. Click the Tools menu again, then Port, and choose the port that says (Arduino/Genuino Uno), as shown in Figure 4.15.

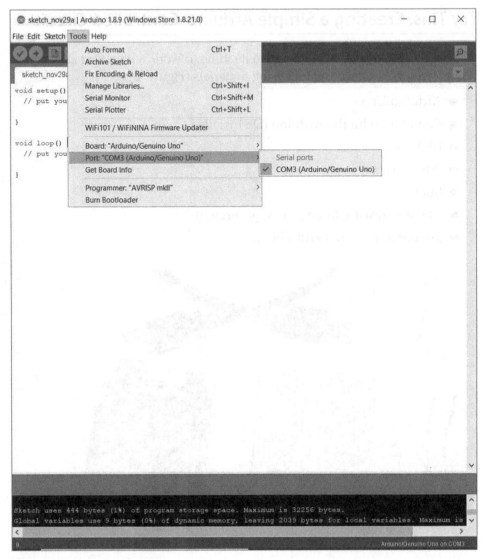

Figure 4.15: Selecting the port

Follow these steps to build the circuit:

1. Insert a black jumper wire into the ground pin (GND) next to pin 13 on the digital input/output side of the board. Connect the other end of the black jumper wire to 1a on the breadboard.

2. Insert a red jumper wire into pin 13 on the digital inputs side of the Arduino board. Connect the other end of the red jumper into the power rail of the breadboard.

3. Insert one end of the 330-ohm resistor into 1d on the breadboard, and insert the other end into 10d on the breadboard.

4. Insert the long lead (+) of the LED into the power rail of the breadboard, and insert the short lead into 10a. Your breadboard should look similar to Figure 4.16.

Figure 4.16: A flashing LED circuit

5. The Arduino IDE has several built-in sketches that are a great place to begin programming. In the Arduino IDE, click File ➪ Examples ➪ 01.Basics ➪ Blink, as shown in Figure 4.17.

Figure 4.17: Choosing the Blink sketch

6. The Blink sketch should open. Click the Verify button (check mark) and wait until the message area says "Done compiling."

7. Click the Upload button (arrow pointing right) next to the Verify button. This will upload your sketch to the Arduino board.

The TX and RX buttons will flash rapidly. If all went well, the program uploaded to the microcontroller, and the program is executed. The LED should be blinking on and off steadily. Ta-da! You made your first Arduino circuit.

What Went Wrong?

If your circuit isn't blinking steadily along, don't despair. Here are some common issues to check.

The most common problem is that the LED leads are in backward. Verify that the long lead of the LED is toward the positive side of the circuit. This is where color-coding jumper wires black to negative and red to positive will help you.

> **NOTE**   Most LEDs follow the convention of long lead to positive and short lead to negative, but you may find some that don't. Use the process that follows to test the LED, but try it with polarity reversed too, in case you have a nonstandard LED. The negative end of a component is known as its *cathode*, and the positive end is the *anode*.

Use a 3V coin-type battery to test your LED, placing the positive to the long lead and negative to the short lead. Is the LED good? (Did it light?) If not, replace it with a good one.

Follow the electricity around from the ground connection on the Arduino board, through the resistor, then the LED, and back to the positive rail of the breadboard. Are any of the connections loose or perhaps one row off?

Is there a break in the red line of the breadboard? Some breadboards allow for two different power inputs by breaking the positive rail somewhere in the middle. Make sure your LED is connected to the same section of the positive power rail as the red jumper wire from the Arduino board.

Verify that you chose the correct board in the IDE. If the Blink sketch opened up a new window, go to that window and follow the previous steps to choose the right communications port.

What Does the Code Mean?

Take a look at the Arduino IDE sketch that you just uploaded to your Arduino board. Two sections to the actual code always need to be there: setup() and loop(). This section is a breakdown of what you see in the Blink sketch.

// indicates that what follows is a comment and not part of the program. This is useful when writing more complex programs to help you remember why you did what you did there. Trust me from experience, you'll want to use those comments.

Most programming languages can be divided into two basic components called *functions* and *variables*. Functions tell the program what to do, and variables hold data that may change from one iteration of the program to another or perhaps one day to the next. Variable names themselves will change from one program to another.

Here is an example line of code: `digitalWrite(LED_BUILTIN, HIGH);`. Notice the way the commands are written: the first word is lowercased, and subsequent words have the first letter capitalized, with no spaces between the words (`digitalWrite`). Mode names (`HIGH`) are in all caps. The variable (`LED_BUILTIN`) doesn't have to be unless it was declared that way. The command is listed first; then comes a bracket followed by the variable name, a comma, then a space, and the mode, value, or state of the variable. At the end of each command line is a semicolon. Yes, capitalization, punctuation, and the order in which objects are entered all matter when writing code. This is known as the code's syntax.

Also notice that both the setup section and the loop section have a curly bracket right after the `setup()` or `loop()` command and again at the end of the section.

Each line of the Blink sketch has a comment after it so you can see what each line does.

Setup

`void setup()` tells the sketch to launch the `setup()` function. Anything in the setup section runs only once when the board is powered on or reset. It is used to set conditions for the program that won't change and to declare or initialize variables. `void` tells the sketch that it doesn't need to return any data from this function to the program.

`pinMode()` in the setup section tells the program how to treat the referenced pin. The syntax is `pinMode(pin, mode)`. In this case, `pinMode(LED_BUILTIN, OUTPUT);` is telling the program that the `LED_BUILTIN` pin, which is pin 13, is set to be an output pin. `pin` can be any of the analog or digital input/output pins. `mode` can be one of three settings: `input`, `output`, or `pull_up`.

The default mode of the pins is input. Not much power is needed to change the state of an input pin, so the pins are not a heavy load on the circuit that they're measuring. Pin 13 is better as an output pin because there is an LED and a resistor built into it already.

Output pins can have a current as high as 40mA, which can be too much for some circuits. Remember, you'll always need a current-limiting resistor in series with an LED to avoid burning it out. The pins on the Arduino can also be damaged if too much current is allowed to flow through them, so a good practice is to always use a resistor on an output pin to limit the current. Pins can also be damaged if too much current or voltage is allowed to be read by them, and in future projects we'll examine how to limit what a pin reads. (As mentioned in Chapter 2, "Electricity: Its Good and Bad Behavior," although the current on a microcontroller circuit can be very small, there must always be a load (or enough resistance) on a circuit to limit the current and keep it from running rampant and causing the conductors to overheat.)

The `pull_up` mode is an input setting that can be used to establish a voltage when there is no input or to act as a NOT gate. A NOT gate changes the input so that low equates to high and high equates to low.

Void

Code in the `void loop()` section repeats continuously until the board is turned off. Again, `void` tells the program that it doesn't have to send any data back to the program to be acted upon, and `loop()` is the part that says, "Keep doing this until I tell you to stop." The program will execute the lines after `loop()` in order, and when it reaches the bottom, it starts right back at `loop()` again.

`digitalWrite()` will send a voltage to a specific pin if the pin is an output pin. The syntax is `digitalWrite(pin, value)`, where the pin is any output pin, and the value can be either HIGH, which is 3.3V or 5V, depending on the particular Arduino board being used, or LOW, which is 0V. In this case, it is told to send a high voltage to the LED on pin 13.

`delay(1000)` tells the program to pause for 1,000ms (milliseconds), equal to 1 second. Program times are always in milliseconds. A program time of 1 millisecond would be 0.001 of a second, and 0.001 times 1,000 equals 1 second.

Next, `digitalWrite()` will send a low voltage to pin 13, so the light goes off and, again, pauses for 1 second.

Try This: Changing Pins

Any of the digital pins on an Arduino board can be used as either input or output, but they must be initialized before you can use them.

1. In the `setup()` section, change `pinMode(LED _ BUILTIN, OUTPUT);` to `pinMode(12, OUTPUT);`.

2. In the `loop()` section, change the first delay to `delay(250);` or another number that makes you happy. Also, change the comment after the // to read 1/4 second, or whatever time you chose.

3. Change the lines that say `digitalWrite(LED _ BUILTIN, HIGH);` and `digitalWrite(LED _ BUILTIN, LOW);` to `digitalWrite(12, HIGH);` and `digitalWrite(12, LOW);`.

Your sketch should look like Figure 4.18.

4. Click the Verify button. If it's all entered correctly, "Done compiling" will appear in the message area. If it's not entered correctly, a message will appear in the message area at the bottom of the screen. Make any necessary adjustments and verify again.

5. Move the red jumper wire on the Arduino from pin 13 to pin 12.

6. Click the Upload button to send the new program to your Arduino microcontroller. The TX/RX lights will flash, and you should see a change in the LED's flash pattern.

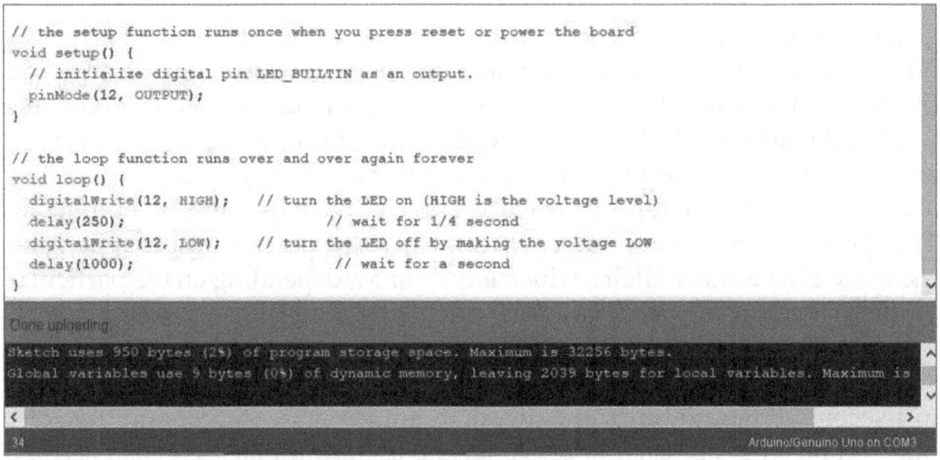

```
// the setup function runs once when you press reset or power the board
void setup() {
  // initialize digital pin LED_BUILTIN as an output.
  pinMode(12, OUTPUT);
}

// the loop function runs over and over again forever
void loop() {
  digitalWrite(12, HIGH);    // turn the LED on (HIGH is the voltage level)
  delay(250);                        // wait for 1/4 second
  digitalWrite(12, LOW);     // turn the LED off by making the voltage LOW
  delay(1000);                       // wait for a second
}
```

Done uploading.
Sketch uses 950 bytes (2%) of program storage space. Maximum is 32256 bytes.
Global variables use 9 bytes (0%) of dynamic memory, leaving 2039 bytes for local variables. Maximum is

34   Arduino/Genuino Uno on COM3

Figure 4.18: Changed Blink sketch

Try This: Creating Arduino Running Lights

Time for some serious fun. Now that you know how to initialize pins and flash LEDs, why not add more?

For this project, you'll need the following materials:

- Arduino Uno
- Computer with the Arduino IDE installed
- USB cable
- Breadboard
- (3) LEDs
- 330-ohm resistor (orange, orange, brown)
- (4) Jumper wires

1. If you just finished the previous project, remove the components and jumper wires to start fresh.

2. On your breadboard, insert one end of a black jumper wire into the negative rail and the other end into the ground pin next to digital pin 13 on your Arduino Uno.

3. Insert one end of a 330-ohm resistor into the ground rail of the breadboard and the other into hole 6a.

4. Insert the negative end of each LED into row 6 using holes 6b, 6c, and 6d, or similar, as long as the negative (short) lead of each LED is in row 6. Place the positive end of one LED in 10b, the next in 11c, and the last in 12d.

5. Insert a jumper wire in the breadboard in hole 10a and the opposite end in pin 12 on the Arduino. Insert the next jumper wire in hole 11a with the opposite end in pin 11 on the Arduino, and a third jumper wire in 12a with the opposite end in pin 10 on the Arduino. For this project, only one resistor is needed because only one LED will be lit at any given time. The resistor is common to all the LEDs in row 6.

6. Open the Arduino IDE. Click the New button and name your sketch Running Lights. Try setting up the program on your own, based on the instructions in this paragraph and the next. If you get stuck, no worries, as the code is shown after the steps. In the `setup()` section of the sketch, initialize pins 10, 11, and 12 to be output pins.

7. In the `loop()` section of the sketch, set up pins 10, 11, and 12 with `digitalWrite(X, HIGH);` and `delay(1000);` and then `digitalWrite(X, LOW);` and `delay(10);`. Replace the x with the actual pin number. The short delay when the light goes LOW (off) will ensure that only one light is on at a time.

7. When the coding is done, click the Verify button. Make any necessary corrections and then verify again. When the code is correct, connect the USB cable and upload the code to the Arduino's microcontroller.

The lights should flash in order, each for a second with no discernable delay between (Figure 4.19). If you're not comfortable yet, here is how the code should look:

```
void setup() {
pinMode(12, OUTPUT);
pinMode(11, OUTPUT);
pinMode(10, OUTPUT);
}
void loop() {
digitalWrite(12, HIGH);
delay(1000);
digitalWrite(12, LOW);
delay(10);
digitalWrite(11, HIGH);
delay(1000);
digitalWrite(11, LOW);
delay(10);
```

```
digitalWrite(10, HIGH);
delay(1000);
digitalWrite(10, LOW);
delay(10);
}
```

Figure 4.19: Running Lights in action

Try This: Adding a Switch to Your Circuit

Perhaps there will be a time when you want to leave all the wires connected but would like to turn off the lights, or whatever device you are running with the Arduino and circuit board. It would be problematic to disconnect the power each time, so a switch is needed. This little project shows you how to add a switch. Adding a switch is one of the first projects that many new Arduino users learn. For this project, you'll need the following materials:

- The running lights project completed and working
- PC with the Arduino IDE installed
- USB cable

- Push-button (momentary) switch
- 10k-ohm resistor (brown, black, orange)
- Several jumper wires

The objective is to set up the switch as an input to the Arduino so that when the switch is in an off state, the Arduino will output voltage low (zero) and the light circuit will not run.

1. Unplug the Arduino from your computer and/or other power source.

2. Open the program created for Running Lights, as it's easier to modify that one than to start over. Save the file with a different name such as Switch.

3. The first entry needed is to set the input pin for the switch. This line of code will be added before `void setup()`. You'll use pin 8, just because it's close to the other pins, but any input/output pin would work. Again, you want to steer away from using pin 13 because it is associated with an internal resistor and LED. The code for setting the pin is `const int switchIn = 8;`. In this code, `const` tells the program that the variable is not going to change. `int` tells the program that the value is an integer, a whole number. `switchIn` is what the variable is being named, and the value is 8, for pin 8.

4. Another entry is needed to set the beginning *value* for the switch. The state of the switch will change from off to on, so it is not a constant. We *want* it to change when the button is pressed. `int switchValue = 0;` sets the initial value of the switch to logic low. This entry also goes before the `void setup()` section.

5. `pinMode` has already been set for the three LEDs in the `void setup()` section. It needs to be set for the `switchIn` pin as well. Add another line in with the other `pinMode` settings that says `pinMode(switchIn, INPUT);` because the switch will provide input to the Arduino board.

6. Now for the fun part. In the `void loop()` section, the first task is to make the program read the present condition of the switch. It's important to use the same variable names that were set up in the previous section. On the first line after `void loop() {` enter `switchValue = digitalRead(switchIn);`. This line tells the program to read the value of the switch. It should be 0 at the moment because that's what was set up in the `int switchValue = 0;` command, but remember that since this line of code is under the `void loop()` section, it will be read each time the program loops back to the beginning.

Now the program needs to see whether the switch has been pressed. For that, an `if` statement is needed. The `if` statement is called a Boolean logic statement because it will equate to either true or false. Arduino also calls an `if` statement a *control structure*. `if` looks at a specified condition and makes a decision based on what it sees: if this, then that, otherwise do something else. In real life, an `if` statement might be "If John is hungry, order pizza, otherwise eat yogurt for dinner." (John doesn't like yogurt.) The syntax for the `if` statement is `if (condition) { //statement(s) }`. If the condition is true, the statements are executed. Note that it uses curly brackets, and you wouldn't put the `//` in. They're just there to tell you that `statement(s)` is a comment, not code. For the "otherwise" part of the equation, the `else` control structure is used.

7. In this circuit, if the button is pressed, its value becomes high. If the value is high, it should turn on the LEDs. Otherwise, it should turn the LEDs off. Here goes: on the line after `switchValue = digitalRead(switchIn);`, enter this:

```
if (switchValue == HIGH) {
```

8. The next lines are already there and will run the running lights and end in a } at the bottom of the code. After the curly bracket, enter `else {` and then tell it to turn the LEDs off with the following entry:

```
digitalWrite(12, LOW);
digitalWrite(11, LOW);
digitalWrite(10, LOW);
}
```

Two curly brackets are needed at the end because one ends the `else` section, and the other ends the `void loop()` section. (One curly bracket should already be there, added automatically when `else {` was entered.)

Whew! That was a lot of entries. Try to make them on your own, and then click the Verify button to ensure that the syntax is correct. If you get stuck or have syntax errors, check your code against the following code:

```
const int switchIn = 8;
int switchValue = 0;

void setup() {
    // put your setup code here, to run once:
pinMode(12, OUTPUT);
pinMode(11, OUTPUT);
pinMode(10, OUTPUT);
pinMode(switchIn, INPUT);
}
void loop() {
    //put code here to run in a loop:
switchValue=digitalRead(switchIn);
```

```
if (switchValue == HIGH) {
digitalWrite(12, HIGH);
delay(1000);
digitalWrite(12, LOW);
delay(10);
digitalWrite(11, HIGH);
delay(1000);
digitalWrite(11, LOW);
delay(10);
digitalWrite(10, HIGH);
delay(1000);
digitalWrite(10, LOW);
delay(10);
} else {
digitalWrite(12, LOW);
digitalWrite(11, LOW);
digitalWrite(10, LOW);
}
}
```

Next, the circuit board needs to be wired to use the switch.

9. There should already be a black jumper from the ground pin next to pin 13 plugged into the ground rail. Verify that it's there.

10. Run a red jumper wire from the 5V pin in the power section of the Arduino board over to the power rail of the breadboard.

11. Verify that the three LEDs are still connected to the negative (ground) rail through the 330-ohm resistor. The other end (long lead) of each LED should still be connected to the respective pins via jumper wires.

12. Place the button across the dip in your breadboard. Pins 20e and 22e and 20f and 22f are the ones used in the picture, so the instructions are based on putting the switch there.

13. Push one leg of the 10k-ohm resistor into 20j and the other into 16j.

14. Push one end of a jumper wire into 16f and the other end into the ground rail. Use black or blue for this jumper because it is the negative side of the circuit.

15. Jumper from 22h to 26h, and from 26f to the positive rail.

16. Make one more connection using two jumpers, from 20i to 28i and 28f to pin 8 on the Arduino board (Figure 4.20).

17. Plug the USB cable from the Arduino into the computer, and in the IDE click the Upload button to send your new program to the Arduino board.

Figure 4.20: A pushbutton circuit

18. The LEDs should be off. Press the button and release it. The LEDs should light in order as they did in the previous project. The program is followed, and the LEDs are turned off at the end.

19. Now press and hold the button.

As long as the button is held in, the input state is high, and the LEDs run. When the button is released, the button's input value goes low, and the LEDs turn off at the end of the cycle. This is a momentary button.

The button in this circuit might be switched out for some other type of switch or sensor, such as a water level circuit or a photovoltaic (PV) cell. Let your imagination go to work. Your circuits are your art.

Try This: Using the Serial Monitor

One final tool for this chapter is the Arduino Serial Monitor. The Serial Monitor enables two-way text communication between the user at the computer and the Arduino board. It works on the 0 (RX) and 1 (TX) pins and will work internally with the USB connection to the computer. It is possible to communicate with another device via pins 0 and 1, but they should not be used other than for communication, or the program or devices could become unstable.

The Serial Monitor must be configured in the sketch before it can be used. It has more than 20 functions, but only a few are covered here.

1. Begin with the sketch from the previous project. Save the sketch with a new name such as Time Lapse. Assume that the user wants to capture the amount of time that elapses between one high input and another.

2. First, in the `void setup()` section, it is necessary to tell the Arduino board how many bits per second to use when communicating with the IDE. The transmission rate is called a *baud rate*. We will use 9600. Enter the following code in the `void setup()` section:

    ```
    Serial.begin(9600);
    ```

3. Next, a variable must be established to capture the time. Above the `void setup()` section, initialize your variable. The example variable is simply named seconds with this line of code:

    ```
    long seconds;
    ```

    Long is a type of variable that can be up to 32 bits long, or a value as high as -/+ 2,147,483,647. A short variable exists that can be up to 16 bits long and numbers up to +/- 32,767.

4. Finally, the actual code to capture the time between peak voltages must be inserted after the first LED goes to the high state. That code is as follows:

    ```
    Serial.print("Time =  ");
            Seconds=millis();
            Serial.print(seconds);
    ```

Each time the output goes high, in this case when the button is pressed, the time will be captured, and the high state will be communicated via the serial port to the serial monitor. Notice that `"Time = "` is in quotes. This tells the program to treat what is inside the quotes as text, not a command or data.

`Seconds=millis();` sets the value of the variable to the number of milliseconds that have passed since the board was reset or the program was uploaded. There are four time functions. The `delay()` function was used in the previous project. The other three are `millis()`, `delayMicroseconds()`, and `micros()`. One thousand milliseconds (1,000) equals 1 second, while one million microseconds (1,000,000) would equal 1 second.

The third line, `Serial.print(seconds);`, retrieves and shows the value of the variable, which was updated in the previous line.

Here is the code in its entirety:

```
const int switchIn = 8;
int switchValue = 0;
```

```
long seconds;

void setup() {
pinMode(12, OUTPUT);
pinMode(11, OUTPUT);
pinMode(10, OUTPUT);
pinMode(switchIn, INPUT);
Serial.begin(9600);
}
void loop() {
  switchValue = digitalRead(switchIn);
  if (switchValue == HIGH) {
digitalWrite(12, HIGH);
Serial.print("Time: ");
seconds=millis();
Serial.println(seconds);

delay(1000);
digitalWrite(12, LOW);
delay(10);
digitalWrite(11, HIGH);
delay(1000);
digitalWrite(11, LOW);
delay(10);
digitalWrite(10, HIGH);
delay(1000);
digitalWrite(10, LOW);
delay(10);
}
else {
  digitalWrite(12, LOW);
  digitalWrite(11, LOW);
  digitalWrite(10, LOW);
}}
```

5. Verify and upload the code to the Arduino board.

6. Now to see the results, the Serial Monitor must be launched. The Serial Monitor can be launched in three ways. While in the IDE, press Ctrl-Shift-M, click the Tools menu on the menu bar, and choose Serial Monitor, or click the magnifying glass icon on the far right of the taskbar.

7. With the Serial Monitor open, press the Reset button on the Arduino, and then press the button on the breadboard.

The Serial Monitor should display the milliseconds between pressing the two buttons. Press the breadboard button again. Again, the time is displayed (Figure 4.21). If this were a real-life sensor, the data could be uploaded to a database to be manipulated further. Remember to save changes to the time-lapse file before closing it.

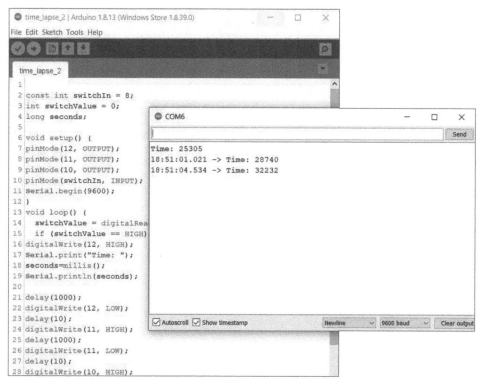

Figure 4.21: Serial Monitor with data

NOTE   The baud rate isn't actually a "bits per second" rate. It refers instead to the number of times per second that the value changes between binary 1 and binary 0. If you're unfortunate enough to need to configure an old-school modem, you'll see baud rate there as well. Baud rate and bits per second (bps) are both measures of the speed of serial (one bit at a time) communications. The term *bit*, by the way, refers to a single binary digit.

CHAPTER

5

Dim the Lights

"We ask ourselves, 'Who am I to be brilliant, gorgeous, talented, fabulous?' Actually, who are you not to be?"

—Marianne Williamson

It brings me great joy when I see other women and girls who let their brilliance shine through and who are not afraid to be proud of who they are. In grade school, I failed a test on purpose once because I had been teased for being too smart. Luckily for me, my teacher knew what was going on. She told my mom, and together they convinced me to ignore anyone who teased me about it and just be myself. That moment in time made all the difference in my life. Be yourself. You are unique, and the world needs your unique talents. Go. Create. Be brilliant. Be your fabulous self.

In this chapter, you'll learn how changes in resistance can "dim the lights." However, this chapter examines much more than that, it examines how to measure electricity. There will be times when something isn't working just right and you'll need to figure out why. There are two important parts to measuring: the first is knowing what to expect the value to be, and the second is knowing how to measure it.

Specific prefixes are used to indicate values when working with electronics. It's important to understand these prefixes, because there's a big difference between 1MV (1 million volts) and 1mV (1/1000 of a volt). There are higher and lower units, but these are the prefixes you're most likely to see:

SYMBOL	UNIT	MULTIPLIER
G	Giga	1,000,000,000.
M	Mega	1,000,000.
k	Kilo	1,000.
	Base unit (A, V, Ω)	1.
m	Milli	0.001
μ	Micro	0.000 001
n	Nano	0.000 000 001
p	Pico	0.000 000 000 001

For example, if a drawing or parts list says a 2.2k-ohm resistor is needed, you would multiply 2.2 by 1,000 and find that the value of the resistor is 2,200 ohms. After working with circuits for a short time, the units will become second nature.

Using a Multimeter

Chapter 2, "Electricity: Its Good and Bad Behavior," explained Ohm's law and the relationship between current, voltage, and resistance. Here's a quick refresher: $E = IR$, meaning that voltage will equal the current in a circuit multiplied by the circuit's resistance. If the voltage applied to a circuit remains constant but the resistance is lowered, the current must go up. It's simple mathematics. Remembering this fact is important, because too much current can destroy devices and cause conductors to overheat and potentially burst into flame.

In purest terms, an ohmmeter measures resistance, a voltmeter measures voltage, and an ammeter measures current. Unless there is a specific reason such as extremely high values being measured, a device called a *multimeter* is used to measure all three of those electrical characteristics. The multimeter combines all three meters into one device, and most multimeters will have other features as well. Figure 5.1 shows a typical inexpensive digital multimeter.

Multimeters can be found online or in local hardware stores and can range from a few dollars to several hundred dollars. I've bought a couple $5 meters that quit working within a week and $20 meters that lasted for years. The expensive meters, which can be hundreds or even thousands of dollars, have features such as audible warnings if the settings and the probe positions don't match, auto-ranging to make setting the meter foolproof, and high accuracy, but for most projects an inexpensive meter should be fine.

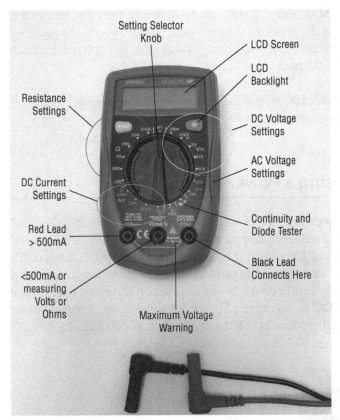

Figure 5.1: A typical multimeter

With all the different features multimeters offer, what should you look for? All meters should have a fuse inside that will protect you if too much current flows through the meter. It's important to know the maximum current and voltage your meter can handle to avoid injury or damaging the meter. Meters come with a red probe and a black probe. Probes are the parts that touch the circuit and are sometimes called *leads*. Most meters will measure both direct current (DC) and alternating current (AC), but they may have different maximum values. A meter with its own stand is important, too, so both your hands can be free to take the measurement and you can still see the liquid crystal display (LCD). At a bare minimum, you'll want a meter that can measure DC amps, DC volts, resistance, and continuity.

The meter's black lead will be plugged into a COM (common) port, which you'll connect to the negative side of the circuit or device being measured. There may be more than one red port for the red lead to be plugged into. Typically, there is a port for measuring resistance and voltage and a different port for measuring current, or different ports depending on how much current you're measuring. Remember, the red probe may have to change positions, but the black probe will always stay in the COM port.

WARNING Be sure to read the manual that comes with your multimeter so that you will know its limits, its symbols, and how to use it safely.

Try This: Repurposing a Power Supply

Who doesn't have a drawer full of old transformers from electronic devices? Those old transformers are likely DC power supplies that can be repurposed into a handy power supply for a circuit board or some small DC device. It's important to know what voltage and maximum current they can supply, and that information is usually found on a label or imprint on the plug. Also, make sure that it is, indeed, a DC transformer. Figure 5.2 shows one of these power transformers. Notice that the input is listed as 120VAC and the output is 12VDC and 300mA.

Figure 5.2: AC-to-DC power transformer

Connecting positive power to the negative side of a circuit could be disastrous, so the other important factor to determine is the *polarity* of the DC power supply. To figure out its polarity you use a multimeter.

For this project, you need the following materials:

■ Multimeter

■ AC-to-DC power transformer

■ Wire strippers

And possibly the following:

■ 22-gauge wire

■ Soldering iron

■ Electrical tape

■ Helping hands

1. Ensure that the power supply is unplugged from the wall outlet, and then cut the plug off the end of the wire.

2. Use wire strippers to remove 1/4 inch of insulation off the ends of the wire. Secure the ends of the wire in a way that they won't move and you can touch them with the probe but they aren't touching anything else. See Figure 5.3, which shows the wires being held by a "helping hands." Don't simply set them on the table because the electricity could travel across the table and injure you or someone else, or damage your equipment. Make sure the wires are far enough apart that the meter leads won't accidentally touch together when you're taking the measurement.

Figure 5.3: Testing polarity

> **WARNING**   Never allow the ends of the stripped wires to contact each other or you when the power supply is plugged in. Doing so could cause a short and harm you, damage the power supply, or both.

3. Set the meter to a DC voltage value slightly higher than the maximum voltage printed on the power supply. For example, if the power supply says it provides 9VDC, set the meter to 20VDC. DC voltage on a meter is usually indicated by a straight line, while AC is a wavy line. Some meters have the two settings combined. The meter shown in Figure 5.1 has separate settings for AC and DC voltage.

4. Verify that the meter leads are plugged into the correct ports on the meter to measure DC voltage.

> **WARNING**   Keep meter leads apart when measuring voltage or current to avoid shorting them out and to avoid damage to the meter or danger to you!

5. Plug the power supply into the wall outlet.

6. Briefly touch one meter lead to each end of the stripped wire.

7. Observe the meter reading. If the value is positive, then the red lead is connected to the positive power, and the black lead is connected to negative. If the reading is negative, it simply means that the leads are reversed and the black lead is connected to positive.

8. Once you determine which wire is positive, unplug the power transformer, then mark the positive wire in some way such as by placing a piece of red electrical tape or red marker on the wire.

If the stripped wires are stranded rather than solid, you may want to solder a small stripped piece of 22-gauge wire onto the end of the stranded wire. If you do, be sure to cover the joint with electrical tape or shrink-wrap. (See the "Soldering, Perfboards, and Shrink Tubing" section of this chapter for more information about soldering and shrink-wrapping. Chapter 16, "Transformers and Power Distribution," will explain how transformers work.)

Congratulations! You repurposed an otherwise useless old power supply transformer into a power supply for your breadboard or other projects!

Measuring Voltage, Current, and Resistance

A multimeter needs to be connected to a circuit in different ways to measure voltage, current, and resistance. Connecting the meter properly is important to ensure your safety and to avoid damaging the meter (Figure 5.4).

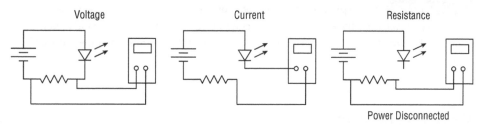

Figure 5.4: Taking measurements

Measuring Voltage

In the previous activity, voltage was being measured. Whenever voltage is measured across a power source, it is called the *voltage rise*. Conversely, *voltage drop* is the difference in potential pressure from one side of a circuit component to the other. Referring to the water-in-the-hose analogy from Chapter 2, the kink in the hose would be a resistance to the flow of water, much like a resistance in an electrical circuit. Assume the hose is partially kinked, so some water still flowed through. If we could measure the pounds per square inch (PSI) in the hose before the kink and again after the kink, the difference between the two would be the pressure that was lost. In electrical circuits, this lost pressure is called *voltage drop*.

To measure voltage drop, the meter needs to be connected *in parallel to* the component being measured, and the power must be on. In parallel means that a meter lead will be on each side of the component that's being measured. The meter should always be set to a value higher than what is expected. If you are unsure what the voltage reading will be, it's best to set the meter at the highest voltage and turn it down until a reading appears.

Measuring Current

Current in an electrical circuit is like the volume of water in our hose. To measure the current, the meter needs to become part of the circuit being measured. This means the meter is connected *in series with* the circuit components. This is a situation where leads with clips are handy. First, with the power off, break the circuit, and then connect the meter across the break. As with voltage, make sure the meter is set correctly and at a higher value than you'd expect. You can always turn it down if you don't get a reading.

On the meter, look for the letter *A*, which indicates amperes. Also pay attention to whether the setting is for DC or AC. Some meters will measure only DC current. Most likely, you'll need to move the red lead from the port for voltage and resistance to a different port for current, and there may even be another port for current over a certain amperage, depending on the meter. Once your meter is set and in place, turn the circuit on and read the current.

> **WARNING**   When preparing to measure current, ensure that the power is off before you start! You have to break the circuit to insert the meter, so to avoid injury, power the circuit off, break the circuit, and then connect the meter and power the circuit back on.

Measuring Resistance

Resistance is measured *in parallel to* the component like voltage, but the power *must be off*. The best way to measure resistance is with at least one lead of the component disconnected from the circuit, or completely removed.

Again, make sure to set the meter properly and at a value higher than what you expect to read. For example, if measuring a 1k-ohm resistor, set the meter at 2k (2000) ohms.

The meter uses internal resistance and voltage from its own battery to function, so when measuring resistance, it's important to "zero out" the meter. Use the following steps to zero out your meter:

1. Set your meter value.

2. Turn the meter on.

3. Touch the red probe and black probe together. There will likely be a number displayed.

4. Hold the probes together until the meter reads zero. The meter is now adjusted so that its internal values won't change the value of the measurements you take.

If the meter reads OL (overload) when you're measuring resistance, it most likely means that you've exceeded the meter's capacity to measure resistance or that the resistance is infinite.

Continuity

Continuity is used to indicate whether electrons can, in fact, get from one end of what you're measuring to the other. The continuity symbol typically looks like concentric quarter circles. Look for continuity near the bottom right of Figure 5.1.

To test for continuity, turn the meter dial to the continuity symbol; then touch the red lead to one end of a device such as a fuse, and touch the black lead to the device's other end. The meter then sends a small current through the device being tested, and if the current is received by the other probe, an audible tone will be heard. If, for example, you're testing a fuse, use continuity. A tone tells you the fuse is still good. No tone tells you that the fuse is burned out and no electrons can pass, so the fuse needs to be replaced. Testing can also be helpful to determine whether a wire has a break somewhere from one end to the other.

Try This: Dimming the Lights

A *potentiometer* is a special type of resistor (Figure 5.5). Typically, a potentiometer has three connectors on the bottom. Potentiometers come in different shapes and sizes, but their purpose is the same. By turning a knob or something similar, the resistance from one side of the potentiometer to the other can be changed. The middle lead is usually the common, and the two outside leads can go to one or another circuit, although one outside lead can be used if there is only one circuit.

A   B   C

Figure 5.5: Potentiometer

Testing this particular potentiometer, measuring from the common lead (b) to lead (a), the reading is 5.3k ohms. Measuring from point (b) to (c), the reading is 5.8k ohms, which means that the total resistance on this particular potentiometer is 11.1k ohms. If I then use a small screwdriver and turn the dial in the center all the way to one side or the other, the resistance on one side is 11.1k ohms, and on the other it's zero.

Now, let's put the potentiometer to work. For this project, you need the following materials:

- 10k potentiometer
- (2) LEDs
- (2) 330-ohm resistors (orange, orange, brown)
- Breadboard
- 9V battery with a battery snap (or the power supply you created in the last project, if it's 9VDC)
- Jumper wire

1. Use a multimeter to test your potentiometer. Adjust it using the center dial until it has relatively equal resistance on both sides.

2. Insert the potentiometer's common in breadboard connector 15e. The outer leads are in 14g and 16g. The potentiometer spans the dip on the breadboard shown so there is more space to work in. It's fine if it's all on one side of the breadboard, as long as the leads are in different numbered rows.

3. Insert a resistor between 16h and 21h, and insert another between 14h and 9h.

4. The shorter leads of the two LEDs are in 9j and 21j, with the longer leads in the positive rail.

5. Place a jumper between the negative rail and pin 15a.

6. Connect your power source to the positive rail (LED side) and negative rail (potentiometer common side.) See Figure 5.6.

Figure 5.6: Dimming via potentiometer

The voltage will be split between the two sides of the potentiometer in the same ratio as the resistance, so if they are relatively equal, each side will get approximately 4.5V. Turn the dial slowly in one direction or the other. The purpose of the 330-ohm resistors is to prevent an LED from being overpowered and burning out. As you turn the dial, one light will become dimmer, and the other will brighten.

Turn the potentiometer in the opposite direction, and the opposite lights will become dim and bright. If you choose, *disconnect the power* and measure how the resistance changes from one side of the potentiometer to the other. You'll find that the side with a brighter light will have a lower resistance.

This is just one way to "dim the lights."

Try This: Measuring Circuit Values

It's time to measure the other properties of a circuit. First, make a simple circuit like the one shown in Figure 5.7.

For this project, you need the following materials:

- LED
- 330-ohm resistor (orange, orange, brown)
- Breadboard
- 9V battery with a battery snap (or the power supply you created in the previous project, if it's 9V)
- Jumper wire

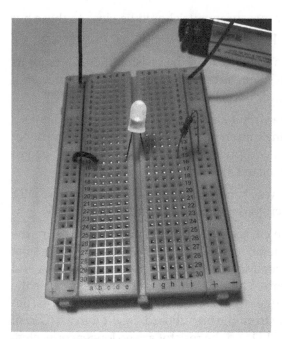

Figure 5.7: Simple circuit

1. Place a jumper in the ground rail and in 15a.

2. Place the short lead of an LED in pin 15e, and place the long lead in pin 15f (crossing over the breadboard dip).

3. Place one lead of a 330-ohm resistor in 15j, and place the other lead in the positive rail.

Chapter 7, demonstrates how to calculate circuit values based on whether the circuit is a series circuit or a parallel circuit. On this circuit, we know that the voltage is approximately 9V and that a typical LED draws 0.02A (20mA) of current. Because we're using a 330-ohm resistor, using Ohm's law (E = IR), the current of 0.02A times the resistance of 330 ohms tells us that the resistor will dissipate approximately 6.6V, leaving 2.4V for the LED. Here's a recap of the expected circuit values:

- **Current:** 0.02A (20mA)

- **Voltage total:** 9V

- **LED voltage drop:** 2.4V

- **Resistor voltage drop:** 6.6V

Now it's time to measure. Your measurements should be close to these but may not be exact, because no component is perfect.

4. First, disconnect the power and one lead of the resistor; then, measure its value. Is it close to 330 ohms? Remember, resistors have a tolerance, so it may be slightly higher or lower than 330 ohms. See Figure 5.8A.

5. Reconnect the resistor in the circuit and connect the power. Adjust your meter to measure VDC at a maximum of 20V. Put the red lead of the meter on the long lead of the LED, and put the black lead on the short lead of the LED. Is the voltage drop near 2.4V? See Figure 5.8B.

6. Leaving your meter set as it was to measure the voltage drop of the LED, measure the voltage drop across the resistor. It should be near 6.6V.

7. Finally, it's time to measure current. Change the meter to measure the amps of direct current. Remember, you must turn the dial and possibly move the red lead of the meter. Most meters have a setting for 20mA and 200mA. Set the meter to 200mA to avoid damaging the meter, because the current might be slightly higher than 20mA. With the power *off*, disconnect the long lead of the LED from the breadboard. Insert it in a different hole, such as 19e, so that the LED and the resistor are no longer connected. Turn the power back on. The LED will not be lit because the circuit isn't complete. Touch the meter's red lead to the resistor, and touch the black lead of the meter to the LED lead. See Figure 5.8C. The LED should light, and your meter should read approximately 20mA.

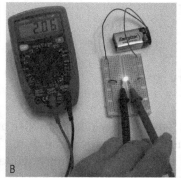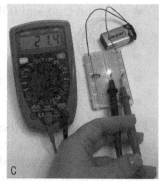

Figure 5.8: Measuring the circuit

Congratulations! You just measured your first circuit.

Using Arduino to Measure Electricity

You may ask yourself why one would want to use an Arduino to measure electricity when a multimeter is so inexpensive and easy to use. What if an Arduino circuit were being used in a field to measure moisture levels? Those levels would somehow have to be converted to a voltage and then reported back somewhere to be of any use. That's where the Arduino comes in handy. Can you imagine sending someone out to 1,000 fields every day to measure the water content? Can you then imagine 1,000 Arduino circuits set up to do the same thing? Data could be reported in real time instead of waiting for the meter readers to all come in from the fields. The cost would become reasonable, because you would need one person to manipulate the data or react to it instead of paying 1,000 people to measure it. Most of the cost would be in the initial setup and making sure the circuit and communications worked as desired. The process to do all of that is complex, and because this is a beginner book, we'll simply consider how an Arduino can be used to take electrical measurements.

Chapter 19, "Connecting Your Circuits to the Cloud" is about communicating via the cloud, so stay tuned.

Try This: Using an Arduino Voltmeter

The following circuit is designed using an Arduino Uno, whose reference value is 5V. If you're using a different Arduino board, the values need to be adjusted accordingly.

For this project, you need the following materials:

- Multimeter
- Breadboard

- Arduino Uno and IDE
- 100k-ohm resistor
- 10k-ohm resistor
- 9V battery with battery snap
- Jumper wires

Chapter 2 introduced the two-resistor voltage divider, and previously in this chapter a potentiometer was used to do the same thing. This project puts the voltage divider to practical use. Dividing the input voltage enables a voltage larger than 5V to be measured with the Arduino board.

Create the following voltage divider circuit on a breadboard. This example uses 100k (brown, black, yellow) and 10k (brown, black, orange) resistors, although higher resistors like 1M and 100k or 10M and 1M would work well, too. (However, higher resisters may require a storage cap to lower the impedance to the ADC.) The higher-value resistor will be connected to the positive power, and the lower-value resistor will be connected to ground with the Arduino pin to measure between the two.

1. Insert one end of the 100k resistor in the power rail and the other end in 5e.
2. Insert one end of the 10k resistor in 5a and the other end in the ground (negative) rail.
3. Insert the black lead of a 9V battery into the ground rail.
4. Insert the red lead of a 9V battery into the + power rail.

Resistors are never perfect; they always have a tolerance, and the larger the resistor, the more important that tolerance is. To get the most accurate voltage readings, it's necessary to figure out the exact ratio between these two imperfect resistors. Use the following steps to determine the ratios on your voltage divider:

1. Set your meter to measure DC voltage. If you're using a 9V battery, the highest voltage should be 9V, so set it to 20V or whatever the next higher value is on your meter.
2. Measure the voltage across both resistors (A to C). This is the voltage being supplied by your battery.
3. Next measure the voltage drop across the higher-value resistor (A to B) and the lower-value resistor (B to C).
4. Write down the values and disconnect the battery. See Figure 5.9.

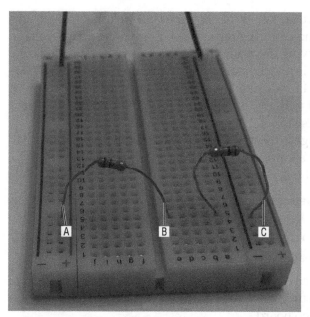

Figure 5.9: Measuring voltage rise

The Arduino will be measuring voltage at the point between the two resistors, so R_2 will be the output voltage. In my example, R_2's value was 1/10th the value of R_1. The voltage measured from point A to point C was 9.05V. The voltage divided by 11 should roughly equal the voltage drop across the second resistor (9.05/11 = 0.822). The measurement taken was 0.82V, so the voltage divider works as expected.

The input voltage (measured from point A to C) divided by the voltage drop across R_2, the smaller resistor, provides a factor that can be used in the IDE to calculate the voltage being measured. In this case, 9.05V / 0.82V = 11.03. Turning the formula around, if the voltage across R_2 is multiplied by the factor, the result is the input voltage. In the example, 0.82V * 11.03 = 9.05V; therefore, whatever the measured value across R_2 is, multiplying it by 11.03 should yield the input voltage.

This particular example allows for measurements up to approximately 50V, but it's better not to go that high. Anything higher t han 55V would damage the input pin (5V * 11.03 = 55.15). It would be better to keep the input voltages no higher than 40VDC, just to be safe.

Remember the comparison between binary values and voltages? By using these resistors, we've changed the top value from 5VDC to 55.15VDC, so the binary numbers available to represent 1V went from 204.8 (1024 / 5) to 18.57 (1024 / 55.15). This means the higher voltage that *can* be measured, the less precise the calculations are for displaying the voltage.

In calculating the voltage read, 5V / 1024 gives the volts per binary digit, and multiplying by the value read from the input pin will give the actual voltage read on that pin. In the following program, the equation is turned around. The value read on the input pin is multiplied by 5 first and then divided by 1,024 because the value of 5 / 1,024 first yields too small a number for the Arduino's processor to interpret, and it will give an answer of zero instead of the correct voltage across the smaller resistor.

The smaller resistor's voltage is then multiplied by the factor found previously (11.03) to arrive at the circuit's input voltage.

All that's left is to attach the Arduino and to create and upload the sketch.

1. Connect a jumper between pin 5c on the breadboard and the Arduino pin A5.

2. Insert another jumper between ground (GND) on the Arduino and the negative side of the breadboard.

3. Insert a jumper into each of the power rails. Color code these for negative and positive to make sure you'll connect them correctly later and avoid damaging your Arduino board.

4. Enter the following code:

```
//Voltmeter >5V

float v_in=0.0; //sets the value of the variable v_in to 0.0
float VRead=0.0; //sets the value of VRead to 0.0
float const factor=11.03; //sets the variable factor to 11.03, and
it won't change.

void setup() {
  Serial.begin(9600); // communicate on serial port

}

void loop() {

  int analogIn=analogRead(A5); // read the reduced voltage on A5
  Serial.println(analogIn); //displays the binary value read from
the pin

  VRead=(analogIn*5)/1024.0; //finds the actual voltage reading
across the smaller resistor
  Serial.println(VRead); //displays the voltage drop across the
smaller resistor
 v_in=VRead*factor; //find the circuit's voltage in
  Serial.print("Input Voltage ");
```

```
Serial.println(v_in);     //displays the circuit's input voltage
>5V.

delay(500);   //measures every 1/2 second

}
```

5. Connect the Arduino to the computer.

6. Verify and upload the code.

7. Open the Serial Monitor.

8. Try your new meter on a battery or a circuit that you know is within range of your meter (not exceeding the maximum voltage!). Touch the lead from the negative rail to the ground or negative side of a circuit or a source. Touch the lead from the positive rail to the positive side of the source or circuit. Note: If reading voltage rise, it's best to do so quickly.

9. Read the voltage on the Serial Monitor.

Ta-da! You did it! If you prefer, comment out the lines `Serial.println(VRead);` and `Serial.println(analogIn);`. Those lines are simply there so you can see what's happening as the microcontroller is reading the voltage and performing the conversion. See Figure 5.10.

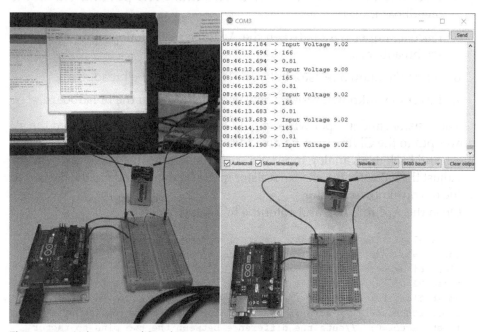

Figure 5.10: Voltmeter and Serial Monitor

Try This: Using an Arduino Ohmmeter

Creating an ohmmeter is much easier than creating a voltmeter. The circuit is so simple that it can fit on a very small breadboard. Creating an ohmmeter uses a voltage divider where one value is known; then Ohm's law is used to calculate the value of the other resistor.

For this project, you need the following materials:

- Multimeter (to check the first resistor)
- Breadboard
- Arduino Uno and IDE
- (2) resistors (at least)
- Jumper wires

1. Measure one resistor using your multimeter so you know its value. Mine measured exactly 500 ohms.
2. Refer to Figure 5.11 for help in placing the jumpers.
3. Place a jumper wire between the Arduino's 5V pin and 2a on the breadboard.
4. Place a jumper wire between an Arduino GND pin and 14a on the breadboard.
5. Place a jumper wire between the Arduino's A1 analog input and 8a on the breadboard.
6. Insert the known resistor between 14c and 8c on the breadboard.
7. Insert the unknown resistor between 8d and 2d on the breadboard.

The resistors and jumper wires should make a complete circuit from the 5V power pin to the GND pin, with an additional jumper where the two resistors meet, and back to an analog input pin so the voltage at that point can be read.

Adjust the value for the rKnown resistor in the following code to the measured value of your known resistor.

Open the Arduino IDE and enter the following code:

```
int analogIn=A1; // sets the analogIn variable to = analog pin A1
int Vin=5;  //sets the voltage in value to 5v
float binary=0;
float V2=0.0; // sets v2 as the second voltage variable
const float rKnown=500;   // sets the value of the known resistor
float rUnknown=0; // sets the variable for the unknown resistor
float factor=0;  //sets the difference between the two pins to factor
```

```
void setup() {
  // put your setup code here, to run once:
Serial.begin(9600);  // communicate on the serial port

}

void loop() {
  // put your main code here, to run repeatedly:
  int analogIn=analogRead(A1);  //read the binary value of voltage for
R2 from pin A1
  Serial.println(analogIn);
  if(analogIn);  //will continue if a voltage was found on pin A1
  {
    binary=analogIn*Vin; //adjusts the voltage to consider the 5V source
    Serial.println(binary);
    V2=(binary)/1024.0;  //changes the binary value to voltage
    Serial.println(V2);
    factor=(Vin/V2)-1; //calculates the ratio between R1 and R2 based on
voltage
    Serial.println(factor);
rUnknown=factor*rKnown; //calculates the resistance of R2
Serial.print("Value of R2 = ");
Serial.println(rUnknown);}
delay(5000);


}
```

Figure 5.11: Arduino ohmmeter

Compile and upload the sketch to the Arduino board, and then launch the Serial Monitor to see the results.

The only `Serial.print` and `Serial.println` lines that are needed are the last two. The other lines that print are for troubleshooting purposes and can be deleted when your circuit is working well.

My R_2 (unknown) resistor was valued at 1200 ohms, so the printed value of 1184 was reasonable with this resistor's 10% tolerance. See Figure 5.12.

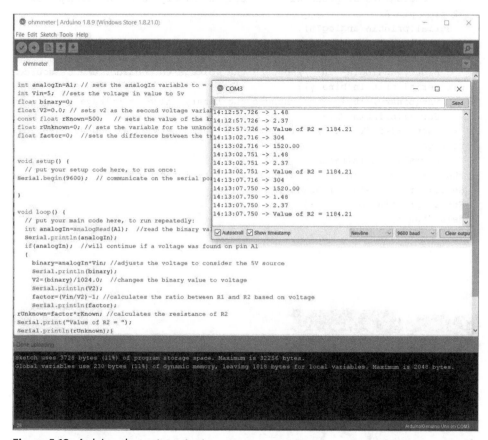

Figure 5.12: Arduino ohmmeter output

Try This: Using an Arduino Ammeter

There are several different ways to build an Arduino ammeter, but since this is a beginner electronics book, we'll keep it as simple as possible. The Arduino Uno can handle a maximum current of about 40mA (0.04A) before the microcontroller chip sustains damage and malfunctions. It's important to keep in mind the maximum limitations of any electronic device that you're using when designing circuits to make sure the equipment stays intact.

Even an inexpensive multimeter can usually handle a few amps of current (100 times what an Arduino can). They're inexpensive and easy to use but not customizable. The beauty of using an Arduino is that you can customize it in multiple ways. The meters we're creating in this chapter aren't necessarily meant for daily use. They're more intended to get you thinking about the creative possibilities. For example, if you need to monitor the current and voltage provided by your water-powered turbine or photovoltaic system, then setting an Arduino system up on a permanent basis might be just the perfect solution.

This ammeter uses a device called a *shunt resistor* (Figure 5.13). The shunt resistor is a small-value resistor placed in series with the circuit. It needs to be small enough so it won't impact the circuit being measured. The resistance value is known, and the voltage drop is measured, so by simply using Ohm's law the current on the circuit can be calculated.

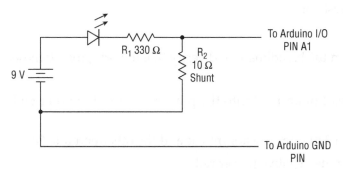

Figure 5.13: A shunt resistor schematic

Looking at the schematic may remind you of a voltage divider. That's essentially what the shunt resistor and the load of the circuit are. The smallest resistor I had on hand was a 10-ohm, 10-watt resistor, but an even smaller resistor would enable greater current measurement. The resistor needs to have a sufficient wattage rating to handle the current. Watts are a measure of power and an indication of the amount of heat produced as electrons are forcing their way through the resistance. The formula for power is $P = IE$, so watts equal current times voltage. A resistor's wattage rating should be greater than the expected watts at that point in a circuit. Power is explained in more detail in the next chapter.

The connection to an analog pin is made between the greater resistance of the circuit and the lesser resistance of the shunt resistor. The analog pin measures the difference between ground and the pin, which is the voltage dropped across the shunt resistor.

Remember that the ADC of the Arduino has 1,024 possible measurements (values 0–1023) spread across 5 volts, and 5 / 1,024 yields 0.00488 volts per binary unit. Ohm's law states that volts divided by resistance equals current, so 0.00488 volts divided by the 10-ohm resistor gives a current of 0.000488 per binary unit.

This, multiplied by 1,024, means that the *maximum* current that can be measured with this resistor is 0.499 A, or 500mA. A lower-value resistor, say 2 ohms, would enable measurement of up to about 2A. This is an extremely simple ammeter, so I recommend keeping what is measured with it on the low side.

For this project, you need the following materials:

- Multimeter
- Breadboard
- Arduino Uno and IDE
- Jumper wires
- 6V lamp or LED
- 330-ohm resistor
- 10-ohm, 10-watt resistor
- 9V battery

To create the circuit on the breadboard and verify that it works, follow these steps:

1. Insert the long lead of an LED into the power rail and the other end into 5b.

2. Insert one end of a 330-ohm resistor into 5c and the other end into 3c.

3. Jumper from 3e to the negative power rail.

4. Attach a 9V battery to the positive and negative power rails.
   The LED should light. Now disconnect the power and remove the jumper from 3e to the negative power rail.

5. Insert the 10-ohm resistor on the other side of the breadboard with one lead in 3h and the other in 20h (or wherever is convenient in another row).

6. Jumper from the second lead of the 10-ohm resistor to the ground rail.

7. Jumper between 3f and 3g.

8. Jumper from the GND pin on the Arduino to the ground rail of the breadboard.

9. Jumper from A1 on the Arduino to 3d on the breadboard. This jumper should be between the two resistors.

Your circuit should look like Figure 5.14.

Figure 5.14: Ammeter circuit

10. Enter the following program into the Arduino IDE:

```
//Ammeter <0.48A

float vShunt=0.0; //sets the value of the variable to 0.0
const int rShunt=10;  //sets the value of the resistor to a
constant
float binary=0.0; //sets the value of binary to 0.0
float current=0.0;
float mA=0.0;


void setup() {
  Serial.begin(9600); // communicate on serial port


}

void loop() {

  binary=analogRead(A1); // read the binary voltage value on A1
```

```
vShunt=(binary*5)/1024; //converts the binary value to a voltage
current=vShunt/rShunt; //calculates the current using Ohm's law
mA=current*1000; //converts amps to milliamps


Serial.print("Current is ");
Serial.print(mA);  //prints the value
Serial.println(" mA");

delay(3000);  //measures every 3 seconds

}
```

11. Compile and upload the sketch to your Arduino board.

12. Launch the Serial Monitor.

13. Plug the 9V battery into the power rail and ground rail. The LED should light, and you should see output similar to Figure 5.15.

Figure 5.15: Ammeter output

When the original circuit (without the shunt resistor) was measured with a meter, the current showed 13.7mA, so there is a small loss due to the shunt resistor. The smaller the shunt resistor, the less it will impact the circuit.

Try This: Using an Arduino Continuity Tester

Most multimeters have a continuity tester. The continuity tester sends out a small current on one probe and measures to see whether it is received on the other probe. If there is continuity, a buzzer sounds so the user knows that electrons were able to get from one end to the other. This is a great way to test a fuse or a wire to make sure it's not broken somewhere in the middle.

This will likely be the simplest Arduino lab possible, as Arduino has a "built-in" continuity tester.

For this project, you need the following materials:

- Arduino Uno and IDE
- Jumper wires
- A small speaker or buzzer (optional)
- A wire to test

That's right! All that is needed is your and 5.20 Arduino board and a few wires.

1. Connect a jumper wire to one of the Arduino's GND pins.

2. Connect another jumper wire to digital I/O pin 7.

3. Enter the following code into the IDE. The lines that say noTone and tone are for the sound, but there's no harm in leaving them there if you don't use a speaker.

```
void setup() {
  // pin 13 has a built-in resistor and LED.
  // we're going to use it as a continuity tester.

pinMode(7, INPUT_PULLUP);  //sets pin 7 as the input
pinMode(13, OUTPUT);  //sets pin 13 as the output
} // the pullup resistor sets the value opposite of what it
normally is.

void loop() {
  noTone(10);  //turns the sound off

int test = digitalRead(7);  // sets the variable named test to the
state of pin 7.  If there is continuity, it will be low
// to get the LED attached to pin 13 to light when there is
continuity, the values must be set counter-intuitively.
```

```
if (test == LOW) {
  digitalWrite(13, HIGH);  //if continuity, pin 13 will light.
  tone(10, 1500);  //play a tone on pin 10 at a frequency of 1500Hz
} else {
  digitalWrite(13, LOW);  //otherwise, it won't light.
  noTone(10);
}

}
```

4. Verify and upload the sketch to the Arduino board. When you touch the two leads from ground and pin 7 together, the LED on the Arduino board connected to pin 13 should light. Touch the leads to each end of a wire or fuse to test for continuity. If the wire or fuse is good, LED 13 will light.

5. To get an audible response from the continuity tester (Figure 5.16), connect a small speaker or piezo buzzer to pin 10 and a ground pin. The speaker should make an audible tone when a positive test for continuity is made. The frequency of the sound can be changed by changing the 1500 in the line tone(10, 1500) to almost any other frequency in Hz. The syntax is tone(pin, frequency).

Figure 5.16: Arduino continuity tester

Try This: Building a Dimmable Arduino Camp Light

I live in the Adirondack Mountains, and being the geek girl that I am, I love to go camping and watch the other campers roll their eyes at all the electronic gear I bring with me to make camping life easier. Some even have the audacity to ask what guy built it all for me. Seriously. Sitting here in the cold New York winter thinking about camping gave me the idea for this lamp/flashlight that could be made brighter or dimmer by the push of a button. First, let's look at the circuit (Figure 5.17).

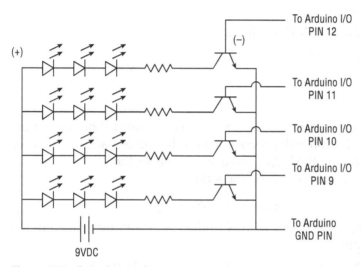

Figure 5.17: Camp lamp schematic

The circuit uses 12 LEDs in four different rows, so a minimum of 3 LEDs will be lit at a time. The project uses four different colors just to make it easier to see, but when building this for a permanent project, you may want to use all super-bright white LEDs. Whatever LEDs are used, it's important to pay attention to the current they draw and the voltage they use. For this project, all LEDs draw a current of 20mA, although the voltages vary slightly by color. A 100-ohm resistor is used in each row. The three LEDs in each row are in series with each other, and the rows are parallel to each other. Each row is connected to a separate output pin on the Arduino. (The math for all of this is explained in Chapter 7).

The circuit also uses transistors, which are explained more in Chapter 9, "Transistors." Transistors are a must in this circuit because the Arduino's maximum current is 40mA, and with all four rows lit, the current on this circuit will be 80mA. The transistors separate the two circuits. Drawing that much current, with a new battery and all four rows lit, the lamp should last a little over six hours.

The transistors have a flat side and a round side. Each lead does something different, so the placement and connection of the leads are important. Remember, this device is protecting your Arduino from the high current, so connect carefully.

The best course of action is to get one row working properly and then add the others in the same manner.

For this project, you need the following:

- Arduino Uno and IDE
- (2) breadboards
- (12) LEDs (four sets of three each)
- (4) 100-ohm resistors
- (4) NPN Transistor BC547B or similar
- Assorted jumper wires
- Push button
- 1200-ohm resistor

For this project, use two different breadboards: a small one to hold the button, considering that the button will probably be on the top or back of whatever container ends up holding the LEDs, and a larger one for the LEDs.

To set up the button breadboard, do the following:

1. Insert the pushbutton across the breadboard dip.
2. Connect a jumper between the row of the button's bottom pin and the Arduino's 5V pin.
3. Connect another jumper between the row of the button's top pin and digital I/O pin 2 on the Arduino.
4. In the same row, connect a 1200-ohm resistor and place the other end in another row.
5. Connect a jumper from that end of the resistor to the Arduino's GND pin. See Figure 5.18.

For the LED circuit, rather than providing step-by-step instructions, which would be tedious with this circuit, we've instead provided an illustration (see Figure 5.20), and the instructions are a bit more general to be consistent with your newfound expertise!

Refer to Figures 5.19 and 5.20 as you work though these instructions. Notice the small red jumper in the middle on the right. A break in the red line on that side of the board can be seen, indicating that the positive power rail is broken there. Some breadboards are set up that way to allow for two different input voltages on the same board. For our purposes, place a small jumper from one side to the other so that the power runs all the way up the rail.

Figure 5.18: A button circuit

Figure 5.19: Breadboard configuration, one row

From the top, the blue jumper connects the power rail to the longer lead of the first LED. The shorter lead is inserted, skipping two rows, and the longer lead of the next LED is in the same row as the shorter lead of the previous one. Continue this pattern of long to short. Figure 5.19 should help you place the LEDs.

Insert one lead of the 100-ohm resistor into the same row as the last LED lead, and insert the other end a couple rows down on the breadboard. This might be a good time to check that your LEDs are oriented properly and working. Plug the positive lead from the battery into the power rail and touch the negative battery lead to the outside (bottom) resistor lead. The LEDs should all light. If they don't, troubleshoot this before going any further.

Insert the transistor so that the flat side is away from the power rail and the first lead is in the same row with the resistor. Connect a jumper wire from the middle lead of the transistor to digital I/O pin 12 on the Arduino board. Finally, connect a jumper wire from the bottom (third) lead of the transistor to the ground rail of the breadboard.

The Arduino and breadboard need to share a common ground, so connect a jumper from the ground rail of the breadboard to a GND pin on the Arduino.

It's time to plug in the Arduino and create the sketch. Remember, the IDE is essentially a text editor, and you can copy and paste rather than retype every single line. Copy/paste saves you time when entering the void loop () section.

```
const int buttonIn = 2;  //sets pin 2 as the one connected to the button
int buttonState = 0;   //starts the button state as off
int count = 0;        // counts the number of times the button has been
pressed


void setup() {
  // put your setup code here, to run once:

pinMode(12, OUTPUT);  //these 4 rows establish the rows connected to
LEDs
pinMode(11, OUTPUT);
pinMode(10, OUTPUT);
pinMode(9, OUTPUT);
}

void loop() {
  // put your main code here, to run repeatedly:

buttonState=digitalRead(buttonIn);
if (buttonState == HIGH) {  //reads if button was pressed
  count++;                   //if yes, increments count by 1
  delay(300);              // waits 300ms
}
if (count == 5) count = 0; //resets the counter to 0 after 5 times

if (count == 4) {            // for each of these sections, as the
```

```
    digitalWrite(12, HIGH);      // count goes down, the number of LED
    digitalWrite(11, HIGH);      // rows that are lit go down.
    digitalWrite(10, HIGH);      //  all pins are high (on) here.
    digitalWrite(9, HIGH);

    }

if (count == 3) {                // if the button has been pressed 3x, only
    digitalWrite(12, LOW);       // 3 rows will light.  Copy and paste the
    digitalWrite(11, HIGH);      // section above, then change one row to LOW
    digitalWrite(10, HIGH);      // and the button count to 3
    digitalWrite(9, HIGH);

    }
    if (count == 2) {            // again the copy and paste, change the
    digitalWrite(12, LOW);       // count to two, and make two pins LOW
    digitalWrite(11, LOW);
    digitalWrite(10, HIGH);
    digitalWrite(9, HIGH);

    }
    if (count == 1) {            // button press count goes to 1
    digitalWrite(12, LOW);
    digitalWrite(11, LOW);
    digitalWrite(10, LOW);
    digitalWrite(9, HIGH);       // notice that only one pin is HIGH now

    }
    if (count == 0) {            // when the count is 0
    digitalWrite(12, LOW);
    digitalWrite(11, LOW);
    digitalWrite(10, LOW);
    digitalWrite(9, LOW);        // all of the output pins are LOW.
    }

    }                            //double check the syntax &amp; brackets..
```

Verify and upload the sketch. Make sure the 9V battery is plugged into the power rails. If you followed the directions given, the first LED row is connected to output pin 12 via the transistor's middle pin. Output pin 12 won't go HIGH until the fourth button press. The LEDs should turn on at the fourth button press and off on press 5. Test it now. Ta-da! Great job.

Now that one row is working, follow the procedure for the other three rows of LEDs. Caution! Make sure none of the components from one LED string is connected to the same horizontal (numbered) rows as another LED string (see Figure 5.20). You need to use both sides of the breadboard to accomplish this.

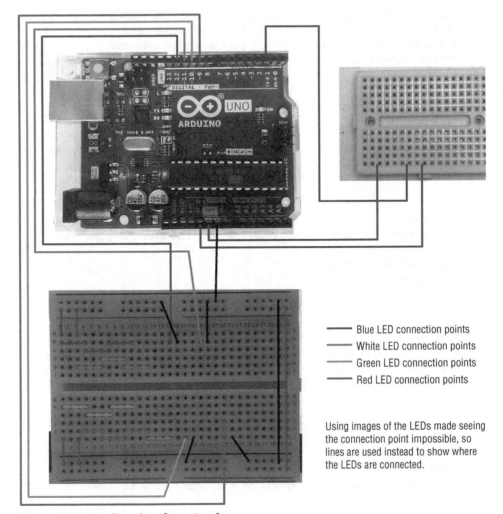

— Blue LED connection points
— White LED connection points
— Green LED connection points
— Red LED connection points

Using images of the LEDs made seeing the connection point impossible, so lines are used instead to show where the LEDs are connected.

Figure 5.20: Breadboard configuration, four rows

Jumper between the positive rail on the right and left, and do the same for the ground rail to make the connections easier.

WARNING   Caution! Make sure none of the components from one LED string is connected to the same horizontal (numbered) rows as another LED string. Doing so would cause a short circuit, and the circuit will not function as it should.

The last three rows will be connected via their transistors to digital I/O pins 11, 10, and 9 on the Arduino board.

This might be a good project to mount permanently on a perfboard. Perfboards provide a more permanent connection than a breadboard, and their use is explained in the next section of this chapter. While you're at it, why not enclose your Arduino and breadboard in a waterproof container, mounting the switch where you can reach it, and take it camping?

Feel free to brag about your finished project by sending pictures to info@ cliffjumpertek.com.

Soldering, Perfboards, and Shrink Tubing

Breadboards are great when designing and troubleshooting circuits, but once the kinks are worked out, it's nice to have a more permanent circuit. Soldering on printed circuit boards (PCBs) and perfboards are the solution.

Soldering

Earlier in this chapter, soldering was mentioned. Soldering uses a hot soldering iron to melt a tin or tin alloy to electrically connect components or parts of a circuit together.

Soldering irons come in various sizes, prices, and configurations. The one I use on a daily basis cost about $20 from an online store. It isn't fancy, but it does the job. Irons with more features can cost up to hundreds of dollars.

To make a good solder joint, both parts that you want to join together need to be hot so they will make a solid metallurgical bond with the solder. Heat can also cause degradation of parts, so it's best to work quickly. A good solder joint shouldn't take more than 5 seconds or so. Often, a component will be soldered to a copper pad. Here's the process:

1. Insert the component lead through the PCB or perfboard so that the component is on the opposite side from the side you'll be soldering.

2. Bend the leads or use helping hands to hold components in place. If possible, trim the leads now so trimming them later won't disturb the solder joint.

3. Place the soldering iron so it touches the pad and component lead.

4. Place the solder on the opposite side of the lead. The solder will melt and wick around the pad.

5. Remove the solder. (Remove the solder before the iron or it may cool too quickly and pull the pad away when you remove it.)

6. Remove the iron.

7. Remember that all of these steps should take only about 5 seconds. It does take practice. See Figure 5.21.

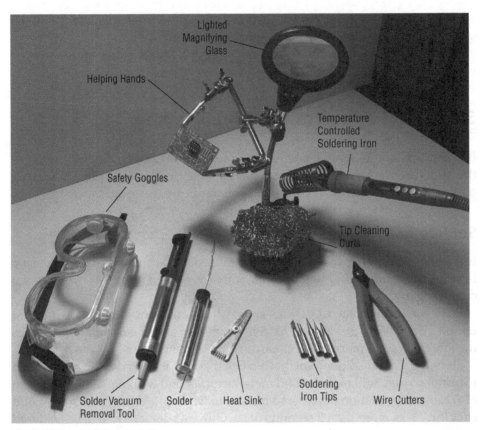

Figure 5.21: A soldering station

It's a great idea to have a carbon filter fan nearby to pull the fumes away when soldering. Most solder includes lead in the mix, such as 60/40, or 60% tin and 40% lead. It usually has a liquid in the center called *flux*, and the vapor you see when soldering is the flux boiling away. The soldering iron likely won't get hot enough to atomize the lead; however, some people are still sensitive to the fumes.

The soldering iron must be kept clean and tinned so it will heat up properly. A damp sponge can be used to clean the tip, but a pot of tip tinner or metal curls are better because they don't cause thermal shock to the tip. The tip should be cleaned and tinned every time you pick the iron up out of its holder. Tinning the tip simply means coating it with a thin coat of solder to assist with heat transfer and to keep the tip from oxidizing. A pot of tip tinner both cleans and tins the tip without thermal shock and will save time on soldering projects.

Helping hands are an essential tool for the soldering workspace. They hold your project at a convenient angle that can be changed, and they free both of your hands so one can hold the solder and the other the soldering iron.

For sensitive components, a small clip called a *heat sink* is placed between the component and the soldering iron to absorb the heat and protect the component.

Transistors and integrated circuit chips are particularly sensitive to heat, as are some other components..

Two common choices for correcting mistakes are solder wick and solder vacuum removal tools. To use solder wick, cut a clean end of wick, spread the end out a bit, and place it on top of the solder to be removed. Press the solder iron into the wick. The solder will melt and be drawn into the copper wick. The process may have to be repeated a few times to get all the solder out. Be sure to remove the wick quickly when removing the iron, because if the solder cools quickly, the wick may pull off the pad as well.

You must be quick to use the solder vacuum removal tool. The vacuum removal tool has a plunger that must be pushed down. There is a button on the side that creates a vacuum when it's pressed, releasing the plunger back to its original position and pulling the molten solder up into the plunger.

To use the vacuum removal tool, press the plunger and then use the soldering iron to melt the solder you want removed. Quickly pull the iron away, place the nozzle of the vacuum against the solder, and press the button; then immediately press the plunger to remove the solder from the vacuum so it won't clog the vacuum.

Both methods take some practice to become proficient. Be patient with yourself while you're learning.

Finally, when through-hole soldering, you'll want to clip the excess leads off to prevent short circuits. Place the curved edge of the clipper toward the board and point the board toward the floor, away from you. Place a finger on the end of the wire being clipped to keep it from flying (Figure 5.22). Safety glasses should be worn when clipping wires.

Perfboards

Perfboards are similar to breadboards because there are rows of holes, but that's where the similarities end. The holes on a perfboard are usually coated with copper for soldering connections, but the holes themselves are not connected together. To create the circuit, one needs to either solder jumper wires in place or, for small distances, make a solder bridge, which is simply solder with no wire.

Some perfboards are stripboards where the rows of holes are connected in strips, and an Exacto knife is used to make separations where necessary.

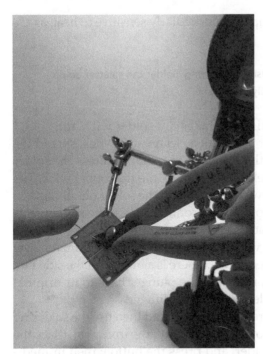

Figure 5.22: Clipping solder leads

Perfboards are a great solution when one or two identical circuits are needed, but if 100 are needed, it's far better to have PCBs made. There are many online sites that will inexpensively print PCBs to your specifications, or you can make them yourself, but that's a subject for another book!

Perfboards come in many shapes and sizes, and Figure 5.23 shows a few examples. The camp light would be an excellent project to put on a perfboard, but it might be wise to practice with a simpler circuit first.

Shrink Tubing

Shrink tubing is a special type of plastic designed to shrink tightly around another object (Figure 5.24). Shrink tubing is sometimes used to waterproof a connection or to simply make a more permanent cover than electrical tape would.

WARNING   Heat guns reach hundreds of degrees and can cause severe burns to skin, hair, or anything in their path. Follow all the manufacturer's directions when using a heat gun.

Figure 5.23: Perfboards

Figure 5.24: A heat gun and shrink tubing

To use shrink tubing, choose a piece of tubing that is closest to the size of what you want to wrap with it. Place the tubing over the end of one component *before* you solder the components together. Solder the components; then slide the shrink tubing over the joint. Hold the solder parts to be shrink-wrapped with long-handled pliers or helping hands to avoid getting burned. Apply heat with a heat gun and watch the tubing closely. The tubing will shrink tightly around the joint. Start with the heat gun a little farther away and then move closer as needed. Use caution! Heat guns reach hundreds of degrees and will melt hair and skin. Do not point a heat gun at anything other than what you intend to heat with it.

Feel the Power

"Personal power is the ability to take action."
—Tony Robbins

All of us can draw on our personal power whenever we choose. Taking action isn't always easy, but taking even a small action toward what you want or the creative project you envision will help you get there. For me, one of the greatest motivators toward action and using my personal power was when someone said, "Girls can't do that." Oh, really!

This chapter is about a different kind of power, and understanding what work and power are from a mechanical and electrical point of view can help you create all sorts of things, too, and maybe keep you from burning the shop down.

Watt's Law and the Power Wheel

Georg Ohm was the German physicist who mathematically explained the relationship between current, voltage, and resistance in the early 1800s, but that wasn't the whole story. Before Georg Ohm, a man named James Watt, a Scottish businessman and inventor, made such an impact on the steam engine industry and our understanding of efficiency and power that the unit for both mechanical power and electrical power, the watt, was named after him.

The formula for power in electricity is *power equals current multiplied by voltage*. The easy way to remember it is that it's easy as pie (P = IE). Power is measured in watts. Knowing how much power is used in electrical circuits is important for several reasons. Power consumption is one of the biggest concerns and most frustrating factors in Internet of Things (IoT) implementations. Power consumption can be a major cost of manufacturing, and too much or too little power can damage components, equipment, and people!

Datasheets

Figure 6.1 shows a snippet of a *datasheet* for a light-emitting diode (LED). A datasheet is a collection of information provided by the manufacturer about a component. The datasheet should always be checked to ensure that a circuit isn't exceeding the maximum current, voltage, or power that the component can handle. It can also provide a myriad of other information that can change, based on the type of component being researched.

ABSOLUTE MAXIMUM RATINGS (T_A = 25°C)

Items	Symbol	Absolute Maximum Rating	Unit
		Red/Amber	
Forward Current	I_F	50 Note1	mA
Peak Forward Current Note2	I_{PF}	200	mA
Reverse Voltage	V_R	5	V
Power Dissipation	P_D	130	mW
Operation Temperature	T_{opr}	-40 ~ +100	°C
Storage Temperature	T_{stg}	-40 ~ +100	°C
Lead Soldering Temperature	T_{sol}	Max. 260°C for 3 sec. max. (3 mm from the base of the epoxy bulb)	
Electrostatic Discharge Classification (MIL-STD-883E)	ESD	Class 2	

Note:
1. For long term performance the drive currents between 10mA and 30mA are recommended. Please contact CREE sales representative for more information on recommended drive conditions.
2. Pulse width ≤0.1 msec, duty ≤1/10.

Figure 6.1: Sample datasheet

The LED described on the datasheet in Figure 6.1 has an absolute maximum power dissipation of 130mW, which is 0.130 watts. The absolute maximum forward current is 50mA (0.05A), but that is only if it is alternating, meaning it is not consistently at that power. In that peak instant, if the current reaches 50mA, what can the maximum voltage on the LED be? Looking at the maximum reverse voltage of 5V, one might assume 5V, but that would be incorrect. If P = IE, then for this component, 0.130 = 0.05E, and 0.130 divided by 0.05 gives us a maximum voltage of 2.6V. I have seen an LED literally explode, sending its top half flying when the maximum power was exceeded. This is why knowing Watt's law as well as Ohm's law is imperative.

The Power Wheel

What if the current and resistance were known, but voltage wasn't? In Ohm's law, E = IR, so in Watt's law IR could be substituted for E, as follows:

P = IE

P = IIR

$P = I^2R$

The fact is that if any two factors among I, E, R, and P are known, the missing factors can be determined by combining Ohm's law and Watt's law, as in the example that we just worked through. If you're in the mood for doing math, feel free to work through all 12 possibilities. If not, the power wheel is a great tool to use to solve these problems. The power wheel combines Watt's law and Ohm's law into 12 simple formulas (Figure 6.2).

The Power Wheel

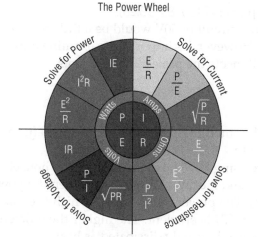

Figure 6.2: The power wheel

Watts and Horsepower

Often a motor will list its power in a unit called *horsepower*. Horsepower was meant to imply the number of horses it would take to perform some task. However, horsepower can be mathematically related to watts and can be calculated to determine how many watts of power are needed to do a certain task.

Horsepower

Why does this matter to us? If a project requires a device such as a mechanical arm to do work for us, then it matters a great deal. Try this example: A mechanical arm needs to move a total weight of 20 pounds to a height of 10 feet, and it needs to complete this action in 5 seconds. How much power will that take?

Work = Force × Distance and is measured in foot-pounds (ft-lbs) or inch-pounds (in-lbs). For our example, that would be 20 lbs × 10 ft, equaling a total of 200 ft-lbs.

Power (in mechanical terms) is work done over time. Power = work/time and is measured in ft-lbs per second or in-lbs per second. The arm has 5 seconds to perform the task, so 200 ft-lbs / 5 seconds equals 40 ft-lbs per second of power.

How big a motor is needed to do this? 1 *horsepower* (hp) is equal to 550 ft-lbs per second, or 40 ft-lbs / 550 = 0.0727 of 1 horsepower. Will a 1/16 hp motor work? No, 1/16 hp is equal to 0.0625 of a hp; however, a 1/8 hp motor will because it's equal to 0.125 hp.

How many watts of power will this motor draw? One horsepower is equal to approximately 746 watts of electrical power. If the motor were 100% efficient, meaning no power was lost from the input to the output, then this 1/8 horsepower motor would need 93.25 watts of power (746 / 8 = 93.25).

If the motor draw was 2A, then a circuit of 50V would be a little more than sufficient (P = IE = 2 * 50 = 100W). However, it's important to examine the datasheet for the motor to make sure it meets the project requirements.

Efficiency

Efficiency is a measure of how well something performs. Calculating efficiency is fairly simple: Power Out/Power In × 100 = Percent of Efficiency.

If in the previous example we're using a 1/8 hp motor running on household voltage (120V in the United States) and it draws 1A of current, then the power in would be 120 watts (1A * 120V). The power out is 1/8 hp, which is equal to 93.25 watts, and (93.25/120) * 100 = 77.7% efficient. The higher the efficiency number, the less power is being wasted, usually dissipated as heat.

Battery Power

The term *battery power* is typically used to mean how long a battery can last. Whether it's your laptop on a long flight or the power source for an IoT device, battery power is an important consideration for any circuit or device.

Battery power is measured in *ampere-hours* (Ah). For example, a battery that supplied 10Ah could run a device drawing 1 ampere of current for 10 hours. A typical 9V battery holds 550mAh (0.550Ah); however, a quick check online can show lithium 9V batteries with a rating of 1200mAh.

The camp light project in Chapter 5, "Dim the Lights," draws 80mA of current when all four rows of lights are lit. When considering the power source, as with any other device, refer to the battery's datasheet for specific information about Ah, current, and operating temperatures.

A typical 9V battery can provide 250mA of current, so the 80mA from the camp light should be just fine. With a typical 550mAh capacity, it should last 6.875 hours (0.550 / 0.08). However, researching battery datasheets shows that the higher the discharge rate, the lower the voltage that can be supplied. So, while a simple calculation shows over 6 hours, it wouldn't burn brightly the whole time and might not last that long. Designing circuits to draw a minimum current is desirable if battery life is important.

Rechargeable batteries are more environmentally friendly, but they tend to have less capacity per charge than a primary (one-time use) cell does.

The Other Resistor Value

Resistors have a rating in ohms and a tolerance value, but they also have a watt rating. A quick search for resistors on `Digikey.com` and `Mouser.com` returned hundreds of thousands of possibilities. The power ratings ranged from 0.01W (1/100th of a watt) to 7kW. Resistors come in various shapes, sizes, and construction materials, and each of those factors influences the maximum power that the resistor can dissipate as heat before it will break down and possibly burst into flame.

Knowing the power that will flow through a circuit's resistors is imperative to avoid problems. Power is calculated the same no matter where it is in a circuit. The power for a given component is equal to the voltage drop of that component multiplied by the current flowing through it. When choosing a resistor, always choose the next largest standard size. The wattage rating of a resistor can always be larger than calculated, because its wattage rating is the maximum power it can safely dissipate, not the power it will use.

For example, a 100-ohm resistor that has a voltage drop of 2.6V on a circuit drawing 20mA would, in theory, need a resistor that is 0.052 watts, calculated as follows: P = IE, so power = 0.02A × 2.6V, which equals 0.052 watts. A factor of at least 10% should be added just to be safe, which would equal 0.062 watts. A 1/8 watt resistor would be fine for this circuit (1/8 = 0.125).

Figure 6.3 shows a few resistors, but there are many more shapes, sizes, and configurations. The resistors across the top are potentiometers (variable resistors). The far-left potentiometer is 100k ohms, and the top right is 1k ohms. The resistors on the bottom range from 1/6W, 100k ohms on the left to 10W, 10 ohms on the right. Size has everything to do with the ability to dissipate power and little to do with the resistance value. The material used and construction of the resistors affect both their resistance and wattage rating.

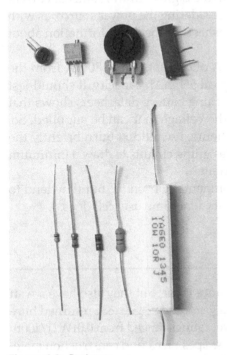

Figure 6.3: Resistors

Wattmeters

As you know by now, power is measured in watts. A *wattmeter* is a handy device used to measure power consumption. In reality, power is never "consumed." The law of conservation of energy teaches us that energy can never be created or destroyed. It merely changes form. In electronic circuits, it is changed between electricity, light, heat, and movement, or waves of energy like sound. Energy being consumed is usually how wattmeters are discussed, so they will be here as well. Be aware that when the word *consumed* is used, we actually mean converted.

Wattmeter

Why would anyone want a wattmeter? If a home is being completely run on solar power and it's located in a place that doesn't get much sunlight, the number of watts being consumed is of great importance. Electrical power is also an important consideration in cost accounting when figuring out the cost to produce whatever is being manufactured. A wattmeter can even be used to monitor the *frequency* of the power being received on an AC system. (Alternating current [AC] will be explained more in future chapters.) The quality of the power being received by a circuit can be extremely important for sensitive devices such as computers and medical equipment.

The formula for power is $P = IE$, so a wattmeter is a device that takes the current and voltage measurements and multiplies them for the user (among other features), saving time and effort. When a wattmeter is combined with some sort of storage, what it measures can be analyzed over the course of time.

Many wattmeters are available commercially, and like other devices can range from less than $20 to thousands of dollars. Also, like other devices, they have limits of current, voltage, and other factors that they can measure without being destroyed. The device on the left in Figure 6.4 is an inexpensive smart plug that includes energy monitoring and advertises being able to analyze a device's power consumption. While the inexpensive plug does provide hours of daily, weekly, and monthly run time, it doesn't give the information in watts. The data and other features for the plug are available via a smartphone app.

Figure 6.4: Power meters

The wattmeter on the right is still inexpensive but has a digital display and can give much more precise information about power quality and consumption.

Power Distribution

While power distribution will be explained in more detail in Chapter 16, "Transformers and Power Distribution," knowing the lingo is important in understanding wattmeters. Power distribution companies have wattmeters in place to see how much power is being used by a customer so they can bill them for it, but the unit they bill in is kWh. What is kWh? kWh stands for kilowatt hours. The prefix *kilo* means 1,000, so a kWh is the use of 1,000 watts of power for an hour. Mathematically, kWh = Watts × Time / 1,000. Electric distribution companies use meters to determine how much power is being used and for how long, which they divide by 1,000, and then bill customers based on the result (Figure 6.5).

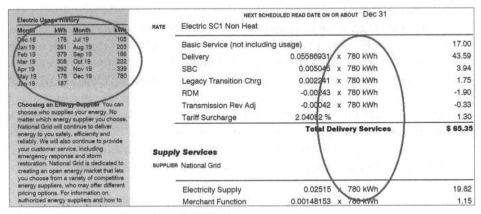

Figure 6.5: Electric supply bill

Try This: Using an Arduino Wattmeter

The beauty of an Arduino wattmeter, as with other Arduino circuits, is that it can be modified to meet a specific purpose, and its data can be uploaded to the cloud and manipulated. Power consumption is one of the biggest challenges of IoT, so monitoring power in preparation for a major deployment of IoT devices is an important consideration. Using an Arduino in this way gives the person or organization that is deploying the device the ability to see what is happening with the device's power consumption on an almost instantaneous basis.

In this project, we introduce the liquid crystal display (LCD) monitor, which can be added to any of the meter projects from the previous chapter to make reading information easier. The plan for this project is to get the LCD screen and then the wattmeter working and then to take a measurement.

For this project, you need the following materials.

For the LCD screen:

- Arduino Uno and IDE
- Jumper wires
- Breadboard
- 10k potentiometer
- LCD TC1602A-21T or similar

For the wattmeter portion:

- Jumper wires
- Breadboard
- 100K-ohm resistor
- 10K-ohm resistor
- 10 Ohm, 10W resistor

For the load:

- 9V battery
- (2) blue LEDs
- 100-ohm resistor
- 47-ohm resistor

Setting Up the LCD Screen

The LCD listed is mounted on a circuit board with built-in resistors and pins to connect it to a breadboard (Figure 6.6). The pins are on the opposite side of the LCD. When you are looking at the board oriented so that the pins are on the top right, pin 1 will be the pin farthest right and pin 16 will be the pin on the far left, toward the middle of the board. The circuit board has a small number 1 and number 16 next to the pins to make it easier.

Pin 1

Figure 6.6: An LCD front and back

A *pinout* is similar to a map of what each pin on a device is. The following table shows the pinout for the LCD:

PIN	WHAT IT'S FOR	SYMBOL
1	Connect to ground (0 volts)	VSS
2	Power supply for logic operations (5V)	VDD
3	Adjusts voltage for brightness on the LCD panel	

(Continues)

Table (*continued*)

PIN	WHAT IT'S FOR	SYMBOL
4	Determines where characters go on the screen. 1: data, 0: instructions (High/Low)	RS
5	Read or write mode (High = Read)	RW
6	Enables data read/write	E
7 through 14	8-bit data bus (8 lines that can be on or off)	D0 through D7
15	Power supply for the backlight	LED+
16	Backlight ground	LED-

Component Connections

Now that you know what all the pins do, proceed with setting up the LCD panel. Refer to Figure 6.7 to help with component placement. This project is a bit more complicated than others that have gone before, so to avoid confusion and for easier troubleshooting, make jumper wires that are just the right length, rather than using premade jumper wires that are too long. The more complicated the circuit is, the more important good organization is. Jumpers should be connected straight or at 90° angles whenever possible, and avoid crossing wires over other wires. Using the ability of the breadboard to connect wires inside will help.

1. Plug the LCD into your breadboard so that each pin is in a different numbered row.

2. Connect the ground pins (1 [VSS], 5 [RW], and 16 [LED-]) to the negative rail of the breadboard. With any device, the ground is usually connected first to give stray electrons a place to go.

3. Insert a 10k potentiometer into the breadboard on the side across the dip from the LCD panel, with each pin in a different numbered row. The potentiometer is used to adjust the backlight on the LCD panel because different viewing angles work better with different contrast levels.

4. Connect the center pin of the potentiometer to pin 3 on the LCD panel.

5. Connect one of the outside pins of the potentiometer to the power rail.

6. Connect pin 2 and pin 15 of the LCD to the power rail.

7. Connect a jumper from the breadboard's ground rail to the GND pin on the Arduino.

8. Connect a jumper from the breadboard's power rail to the 5V pin in the power area of the Arduino.

9. Connect pins 4 and 6 from the LCD to the digital I/O pins 12 and 11 on the Arduino. Connect pins 11, 12, 13, and 14 from the LCD to digital I/O

pins 5, 4, 3, and 2 on the Arduino. These specific pins are being used because they match the sample sketch that will be used to test the LCD; however, any of the digital I/O pins can be used as long as the sketch reflects how the pins have been configured. Pin 4 of the LCD determines where the characters will appear on the screen, and pin 6 tells the LCD panel that the program will be (in this case) writing to it. LCD pins 14, 13, 12, and 11 tell the LCD panel what letter or number to display. Four more data pins are available on the LCD, but they aren't used in this project. Use the following table as a quick reference.

LCD to Arduino

LCD PIN	ARDUINO PIN
4 (RS)	12
6 (Enable)	11
11 (D4)	5
12 (D5)	4
13 (D6)	3
14 (D7)	2

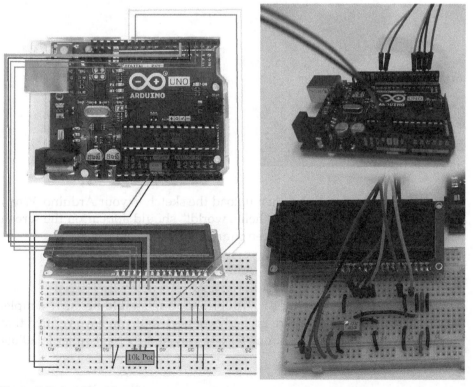

Figure 6.7: An LCD wiring diagram

The Test Sketch

Next comes the sketch. Plug the Arduino into a PC using the USB cable. The Arduino IDE has many example sketches that are free to use. To test the LCD panel, use the HelloWorld sketch. To load the sketch, open the Arduino IDE. Click File ➪ Examples ➪ Liquid Crystal ➪ HelloWorld (Figure 6.8).

Figure 6.8: HelloWorld sketch

Click the Verify button and then upload the sketch to your Arduino. When the built-in LED stops flashing, "hello, world!" should appear on the screen with the number of seconds it's been running below it.

Troubleshooting

If "hello, world!" does not appear, troubleshooting should be fairly simple. First, ensure that all the pins are properly connected. If you have verified that they are, try turning the potentiometer one way and then the other to adjust the brightness of the LCD.

Building the Wattmeter

Once the LCD screen is set up and running, it's time to build the wattmeter. Because a wattmeter combines a voltmeter and ammeter, parts of this circuit should look familiar. It's a simple wattmeter.

Building the Wattmeter

The circuit starts with the familiar voltage divider. The voltage divider section of the circuit is in parallel to the load (refer to Figure 6.9.) This works to measure the voltage on the load circuit, because the voltage on each branch of a parallel circuit is equal to the other parallel branch voltages. The voltage measurement is taken between the two resistors and is the voltage drop across the smaller resistor, from the point between the two resistors to ground.

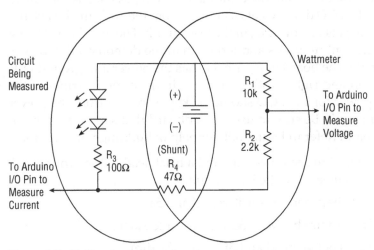

Figure 6.9: Multimeter and circuit

The voltage divider resistors were chosen more for the ratio than their value and happened to be resistors that were available (R_1 = 10k ohms and R_2 = 2.2k ohms). The smaller resistor, divided by the sum of the resistors, produces a factor (0.180328). The maximum voltage output, in this case 5V, that our Arduino can handle as an input, divided by that factor, tells us the maximum voltage that can safely be measured by our circuit. In this case, it's approximately 27.7V. For safety and to avoid damaging the Arduino board, this meter should not be used to measure anything over 24V.

To measure current, a small shunt resistor is generally used in series with the load. The voltage drop is measured across the shunt resistor, and Ohm's law is used to calculate the current. The resistor takes some of the voltage from the circuit, so a smaller resistor is better, generally in the milliohm range. Our

Arduino microcontroller isn't sensitive enough to measure the resulting tiny voltage drop without adding some sort of amplifying circuit, so for now we use Ohm's law to calculate the current using a small resistor that is already part of the circuit. Knowing that the voltage drop across that resistor will be less than 5 volts is again important because a larger voltage drop can damage the Arduino board.

In a parallel circuit, the voltages are equal, but the current on each branch could change, depending on the resistance on that branch. If the resistance on the branches is different, then based on Ohm's law the current will be different. (The differences between parallel and series circuits are explained in Chapter 7, "Series and Parallel Circuits," and using an amplifier to measure small values will be explained in Chapter 12, "Electricity's Changing Forms," when integrated circuits [ICs] are discussed.)

If a full-size breadboard was used to create the LCD circuit, the other half of the board can be used to create the multimeter circuit, as long as the power rails are divided. The LCD display gets its power from the Arduino board, but the wattmeter circuit gets its power from the load circuit. The breadboard being used may have different numbers and letters than the demonstration board, so adjust accordingly. The board in Figure 6.10 has the power and ground rails divided between left and right. Notice the break in the colored stripes of the power rails. The power for the LCD should be kept separate from the power for the wattmeter circuit. Be sure to use a breadboard that separates them or a separate breadboard. Refer to Figure 6.10 for help in building your circuit.

1. Connect a red lead to the power rail and a black lead to the ground rail. These will be connected to the load circuit.

2. Connect R_1 (10k) between the positive rail and 5a.

3. Connect a jumper wire between 5c and Arduino A0 (blue).

4. Connect R_2 (2.2k) between 5d and 5f (across the dip).

5. Connect a jumper wire between the ground rail and Arduino GND (green).

6. Connect a jumper between 5j and the ground rail.

Building the Test Circuit

The circuit being measured also needs to be connected to the meter. For that circuit, a small breadboard is sufficient. The 47-ohm resistor serves as the shunt resistor, although it has been calculated to be part of the circuit so that the value will be large enough to be measured without an amplifier. Be sure to connect carefully. Remember that the Arduino can only have an input up to 5V, so it's important to ensure the voltage being read is low enough to avoid damaging the Arduino board. The voltage must also be measured across a known resistance for the calculation to be accurate.

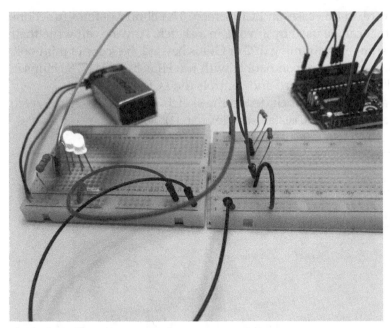

Figure 6.10: Wattmeter connections

1. Connect the 47-ohm resistor to the negative rail and 4k.

2. Connect the 100-ohm resistor to 4i and 6i.

3. Connect a jumper from 4j to pin A4 on the Arduino board.

4. Connect one LED between 6g and 8g and the other between 8h and 10h, making sure to orient them with proper polarity.

5. Connect a jumper from position 10k to the power rail.

6. Connect the red and black leads from the multimeter circuit to the negative and positive rails of the circuit being measured.

Configuring the LCD

Now for the fun part! It's time to tell the Arduino what to do. Ensure that the Arduino board is still connected to the PC via the USB cable. Open the IDE, and then as you read the following paragraphs that explain the process, enter the sections of code that appear there.

Start with a new sketch. Building the parts that support both the LCD and the wattmeter will aid in understanding what they do.

First, `#include <LiquidCrystal.h>` tells the Arduino to include the library file for the LCD in your sketch, but what is a library? A *library* is a program written in C or C++ that tells the Arduino's ATmega chip how to interact with the attached hardware, in this case, an LCD screen. The library defines functions and constants that can be used by anyone in a sketch. Anyone can write their own library files, but programming in C or C++ is beyond the scope of this book. Most LCD 16 × 2 screens are compatible with the Hitachi HD447780 chipset, which is found on the LCD board and controls the LCD.

To use a library, it must be included in the sketch. In the IDE, select Sketch ⇨ Include Library ⇨ Liquid Crystal (see Figure 6.11).

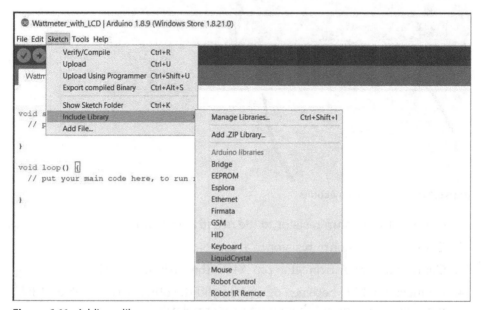

Figure 6.11: Adding a library

Next, it's necessary to identify which pins of the Arduino are matched with the functions of the LCD. The function is LiquidCrystal(). lcd is a variable identifying the LCD. Enter the following lines of code. (#include <LiquidCrystal .h> should have been automatically added when the library was added.) You can omit the // lines because they are comments, but they're helpful to have later when trying to remember what a line is there for.

```
#include <LiquidCrystal.h>  //includes the library in your sketch

// match the Arduino pins with the LCD functions
// refer to the chart in chapter 6.
// RS is LCD pin 4, Enable is LCD pin6, D4, D5, D6, and D7 are LCD pins
5,4,3 &amp; 2 respectively.
// the syntax for the command below is LiquidCrystal(rs, enable, d4, d5,
d6, d7).
```

```
// if RW were not connected to ground, it would be included after rs.
// if 8 bits of data were being used, pins d0, d1, d2 and d3 would go
before pin d4.

LiquidCrystal lcd(12,11,5,4,3,2);


void setup() {

  // specify the dimensions (columns and rows) of the LCD
  lcd.begin(16,2);
  lcd.print("Wattmeter");

delay(3000); // keeps Wattmeter on the screen for 3 seconds
lcd.clear(); //clears the screen and moves the curson to home position
(row1, col1)

}

void loop() {
  // put your main code here, to run repeatedly:

}
```

Click Verify and then upload the sketch to the Arduino to ensure that so far all is working. "Wattmeter" should appear on the LCD screen for three seconds after uploading the sketch (see Figure 6.12).

Figure 6.12: Wattmeter on an LCD

Programming the Meter

Variables for the current reading and voltage reading need to be initialized. Enter the following lines in the section before `void setup()`:

```
int Vin = A0; //initializes the variable Vin to be Arduino pin A0
int Cin = A4; //initializes the variable Cin to be Arduino pin A2
```

A0 is the input pin that the voltage is read from (`Vin`), and A4 is the pin whose voltage reading will be used to calculate the current (`Cin`).

In the `void loop()` section, tell the sketch to read the value for the `Vin` and `Cin` pins.

```
// read the values for voltage and current
   float Vin = analogRead(Vin);
   float Cin = analogRead(Cin);
```

Determining Circuit Voltage

In the IDE, enter the lines to calculate and display the voltage values. Replace the multipliers, if necessary, with values calculated based on your actual resistor values. To find the multiplier for voltage, add the values of R_1 and R_2. Divide that value by R_2. If 10k and 2.2k resistors were used, 12.2k divided by 2.2k gives you a factor of 5.5454 repeating. The resulting number multiplied by the value read on R_2 gives the circuit voltage value, after the binary value is adjusted for 27.7 volts total instead of 5. For a more accurate reading, take the actual measurements of the resistors being used and follow the previous instructions, replacing 5.5454 in the code that follows with the factor you calculate. For the voltage reading, enter the following lines before the `void setup()` section:

```
Float voltage=0.0; //sets the value of voltage to 0.0
float VRead=0.0; // sets the value of VRead to 0.0
float const vFactor=5.5454; //sets the value to multiply R2's voltage by
to get the total
```

Enter the following lines in the `void loop()` section:

```
VRead=(Vin*5)/1024.0; //finds the voltage read across R2
 voltage=VRead*vFactor; //finds the actual voltage
 lcd.print("Voltage = "); //formats output
 lcd.print(voltage); //displays the voltage
  delay(3000); //waits 3 seconds
 lcd.clear(); //clears the screen and returns to home position
```

Determining Circuit Current

In the IDE, enter the lines to calculate and display the current values. Replace the multipliers, if necessary, based on your actual resistor values. To calculate the current, enter the following lines before the void setup() section:

```
Float current=0.0;
float C=0.0; // sets the value of C to 0.0
float const R3=47.0; //sets the value of the R3 to 47 ohms.
```

Enter the following lines in the void loop() section:

```
C=(Cin/1024); //finds the voltage drop on the shunt resistor
 current=C/R3; //finds the current using Ohm's Law
lcd.print("Current = ");
 lcd.print(current);
lcd.print(" A"); //shows that the value is displayed in amps.
  delay(3000);
  lcd.clear();
```

Determining Circuit Power

All that is left to do is to use Watt's law to determine the power converted by the circuit. Add the following line before the void setup():

```
float power=0.0; // sets the power to zero
```

Add the following lines to the void loop() section:

```
power=current*voltage;
   lcd.print("Watts = ");
 lcd.print(power);

 delay(5000);
lcd.clear();
```

The current could also be displayed in mA, by multiplying the value by 1,000 and changing the display to read mA. However, it's important to remember that it must be in the base unit (amps) for calculations of either Ohm's law or Watt's law to be correct.

Click the Verify button and upload the sketch to the Arduino board. Your meter should display "Wattmeter" and then the voltage, current, and power on the circuit, as in Figure 6.13.

Figure 6.13: A completed meter and circuit

Troubleshooting

To aid in troubleshooting, the following listing shows the code from the IDE in its entirety, and Figure 6.14 shows the pin connections for the meter and the circuit being measured.

```
#include <LiquidCrystal.h>  //includes the library in your sketch

// match the Arduino pins with the LCD functions
// refer to the chart in chapter 6.
// RS is LCD pin 4, Enable is LCD pin6, D4, D5, D6, and D7 are LCD pins
5,4,3 &amp; 2 respectively.
// the syntax for the command below is LiquidCrystal(rs, enable, d4, d5,
d6, d7).
// if RW were not connected to ground, it would be included after rs.
// if 8 bits of data were being used, pins d0, d1, d2 and d3 would go
before pin d4.

LiquidCrystal lcd(12,11,5,4,3,2);
int Vin = A0;  // initializes the variable Vin to be Arduino pin A0
int Cin = A4;  // initializes the variable Cin to be Arduino pin A4


float voltage=0.0;
float VRead=0.0; // sets the value of VRead to 0.0
float const vFactor=5.5454; //sets the value to multiply R2's voltage by
to get the total

float current=0.0;
float C=0.0; // sets the value of C c to 0.0
```

```
float const R3=47.0; //sets the value of the R3 to 47 ohms.

float power=0.0; //sets the power to zero.


void setup() {

  // specify the dimensions (columns and rows) of the LCD
  lcd.begin(16,2);
  lcd.print("Wattmeter");

  delay(3000); //delays 3 seconds with Wattmeter on screen
  lcd.clear();  // clears the screen and moves the cursor to home
position (row1,col1)


}

void loop() {
  // read the values for voltage and current
  float Vin = analogRead(Vin);
  float Cin = analogRead(Cin);

 VRead=(Vin*5)/1024.0; //finds the voltage read across R2
 voltage=VRead*vFactor; //finds the actual voltage
 lcd.print("Voltage = "); //formats output
 lcd.print(voltage); //displays the voltage
 delay(3000); // waits 3 seconds
 lcd.clear(); //clears the screen and returns to home position


C=(Cin/1024); //finds the voltage drop on the shunt resistor
  current=C/R3; //finds the current using Ohm's Law
  lcd.print("Current = ");
 lcd.print(current);
 lcd.print(" A"); //shows that the value is displayed in amps.
  delay(3000);
 lcd.clear();


 power=current*voltage;
   lcd.print("Watts = ");
 lcd.print(power);

 delay(5000);
 lcd.clear();



}
```

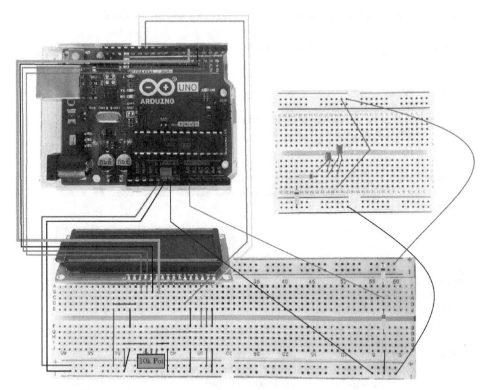

Figure 6.14: Meter and circuit connections

Remember that a common error with LEDs is improper orientation. Make sure that the long leg of the LED is toward the circuit's positive side.

Use a meter to measure the circuit, and compare the values to your wattmeter readings. Remember, if stated values rather than measured values for the resistors were used, the readings may be a bit off because of the resistor tolerances.

For a video of the circuit, visit `cliffjumpertek.com /Arduino_videos`.

Series and Parallel Circuits

"Two roads diverged in a wood, and I —
I took the one less traveled by,
And that has made all the difference."

—Robert Frost

I've been told by someone close to me that the defining moment of my life was in middle school. I wanted to take shop, but I wasn't allowed to because I'm female. Thankfully things have changed since then, but the path set in motion that year affected the rest of my life. Many times, I've been the only female at a conference or in a meeting, but I am doing what I want to do, and that's what matters to me. Do the work that makes you happy, even if people think you're weird for doing so and even if people put obstacles in your path. Decide you can do it, and you can.

Electricity is somewhat like that. It takes different paths, depending on the type of circuit it's traveling through. We put obstacles (components) in its path to cause it to behave in certain ways. Sometimes, it takes paths that we don't want it to, because a dust bunny, a stray piece of wire, or careless circuit building has caused a short. This chapter is about figuring out just how many of those electrons are going in what direction, based on the type of circuit they're in.

Series, Parallel, and Complex Circuits

Understanding the flow of electricity through a circuit is important because too much current or voltage can damage components on the circuit or cause overheating of wires and components that can cause a fire or sometimes even an explosion. Calculations for circuits are different depending on whether the components on the circuit are in series with each other or parallel to each other, and on the type of component being used. To illustrate the difference between series and parallel, we will use resistance.

All components, even conductors, have some resistance. (We'll ignore the conductor resistance for now.) Resistors themselves are specific devices created to limit the flow of electricity through a circuit. When comparing the differences between devices in series and parallel in this chapter, resistance is used. How the electricity behaves with other devices in series or parallel will be discussed in each chapter about those devices.

The simplest way to explain series and parallel is that in a series circuit, the electricity has only one possible path, while in a parallel circuit, more than one path can be taken.

A complex circuit combines both series and parallel elements (see Figure 7.1).

Referring to the water analogy again, the current in a series circuit is like a garden hose—the water can take only one path. The current at point A is the same as it is at point B. A parallel circuit is more like a river that has a Y in it. When water gets to the Y, it will take one path or another, and the paths aren't necessarily equal. The gallons per hour of water in each section of the Y, when added to the other sections, must equal the gallons per hour of water that reaches the Y.

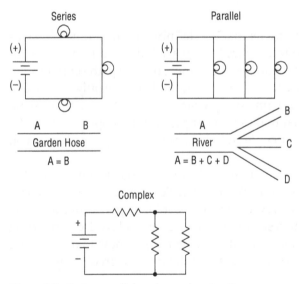

Figure 7.1: Series, parallel, and complex circuits

Try This: Testing Series and Parallel Configurations

The purpose of this project is to illustrate the effect on a circuit if the components are wired in a series configuration or parallel configuration. For this project, you need the following materials:

- Breadboard
- (2) 9V batteries with equal charge remaining
- (6) LEDs of the same color and current (either red or yellow)
- (2) 330-ohm resistors
- 100-ohm resistor
- 6V battery
- (2) 6.3V incandescent lamps with holders
- 22-gauge jumper wire

The LEDs should be identical because different colors, sizes, or intensities require different voltages, and to truly illustrate the difference between series and parallel, equal voltage requirements are needed. Red and yellow LEDs typically require about 2V each. Remember to place the long lead of the LED on the positive side of the circuit.

Use the following instructions to create the two circuits depicted in Figure 7.2. The connections referenced are for the sample board, so adjust the connections to match your board as necessary.

First, here are the steps for the series circuit:

1. Insert a 330-ohm resistor so that one end is in the negative power rail and the other is in 2j.

2. Insert the shorter lead of one LED in 2i, next to the resistor, and the other, longer lead in 5i.

3. Insert the shorter lead of one LED in 5h, and insert the other, longer lead in 8h.

4. Insert the shorter lead of one LED in 8f, and insert the other, longer lead in 11f.

5. Insert a jumper wire between 11i and the positive power rail.

That's it for the series circuit. Now on to the parallel circuit:

6. Insert a jumper wire between 20a and the positive power rail.

7. Insert a 330-ohm resistor between the negative power rail and 23b.

8. Insert the long lead of each LED into a connector in row 20, such as 20b, 20d, and 20e. The LEDs must all be on the same side of the breadboard dip so they will be electrically connected.

9. Insert the shorter lead of each LED into a connector in terminal row 23, such as 23c, 23d, and 23e.

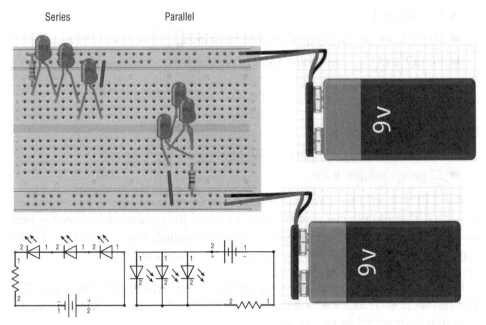

Figure 7.2: LED circuits

10. Insert the power and ground connectors from one 9V battery into one set of power rails, and the other 9V battery into the other set of power rails. See Figure 7.2.

Set the multimeter to measure 200mA of current. Remove the lead of the resistor that is closest to the LEDs for each circuit. Measure the current of each circuit and jot it down.

Now, create the two circuits shown in Figure 7.3 using the incandescent lamps. For the series circuit, connect the lamps to each other; then connect each lamp to an opposite terminal of the 6V battery. Depending on the lamps you have, you may need to strip the ends of lengths of 22-gauge wire or use alligator clips. Observe the brightness of the lamps.

For the parallel circuit, connect each lamp independently to the terminals of the 6V battery as shown.

The measurement on the LED circuits (Figure 7.2) should have shown significantly more current on the parallel circuit than the series circuit.

The incandescent lamps in series (Figure 7.3) should have been much dimmer than the lamps in parallel. Why does this occur? Disconnect the power.

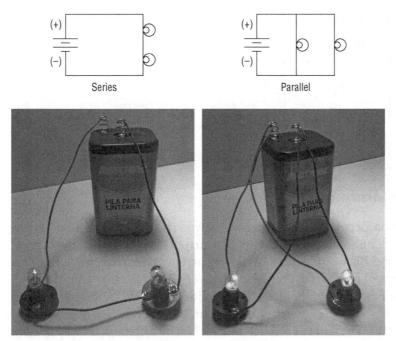

Figure 7.3: Incandescent lamp circuits

Calculating Values in Series and in Parallel

Most of the circuits created so far have been simple series circuits, but many circuits are a combination of series and parallel. Knowing how to calculate values so that the proper wattage resistor or other components can be selected is critical to circuit success. First, let's take a look at calculating current.

Current

Gustav Kirchhoff was a German physicist who, in the 1800s, developed several laws for electrical circuits. Kirchhoff's current law states that the total current entering a junction is exactly equal to the current leaving the same junction. Refer to Figure 7.4, a parallel circuit. The small arrows show the direction of current from the negative side of the source, through the circuit and back to the positive side. Point B, represented by a black dot, is a junction between conductor segments AB, BD, and BC. A *junction* is any point where multiple conductors or components connect to a circuit. A terminal row on a breadboard (a numbered row of five or six holes) can be considered a junction. Thinking of the junction dot on a schematic this way may help in transitioning from a drawing to the breadboard.

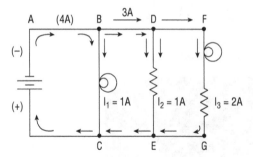

Figure 7.4: A parallel branch current

If this circuit had a current measuring 4A between points A and B (also known as the main branch), according to Kirchhoff's current law, the sum of the current on branch BC and BD would equal 4A. If the current going from point B to C is 1A, then the current going from point B to D must be 3A. If another 1A is dropped down segment DE, then the final 2A is left for segment FG. A simpler way to say this is $I_T = I_1 + I_2 + \ldots I_N$, where N = the number of branches (segments) in the parallel circuit. The current on each branch can be different. Simply adding the current of all the other branches will equal the current of the main branch.

In a series circuit, the current is the same wherever it is measured in the circuit because there is no other branch for the electrons to flow into.

On the parallel circuit project with the three LEDs, because each branch is the same, dividing the main branch current by 3 should give the current of each of the other branches. For example, if the current measured was 24mA on the main branch, the current across each LED would be 8mA. If the series circuit current measured 10mA, then the current across each LED would be 10mA.

A parallel circuit is also known as a *current divider* because there are two or more branches to divide the current across.

Voltage

In a series circuit, the voltage drop depends on the resistance of the component being measured and can change at each component. Voltage drop is easily measured or calculated using Ohm's law when the resistance and current are known.

Kirchhoff's second law, known as Kirchhoff's voltage law, tells us the sum of the voltages in a closed loop is zero. That means that if a meter were used to measure the voltage rise of the source with the meter leads placed, as shown in Figure 7.5, a positive number would result. Perhaps 9V. If the meter leads were moved around the circuit to measure the voltage drop of each resistance without changing the order of the red and black leads, then negative values would result. Adding all of the voltage values in Figure 7.5, including the source voltage, would result in zero.

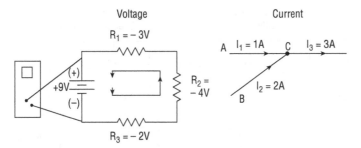

Figure 7.5: Kirchhoff's laws

A simpler way to think of this is that in a series circuit, the sum of the voltage drop of all the components is equal to the source voltage. Two or more resistances in series can be used to divide a voltage. Voltage dividers were introduced in Chapter 2, "Electricity: Its Good and Bad Behavior," and used in the voltage measuring project in Chapter 5, "Dim the Lights."

In a parallel circuit, each branch's voltage is the same as the source.

Resistance

In a series circuit, resistances are calculated similarly to voltages. The sum of the resistance of all the components is equal to the total resistance. If the resistance of a component is unknown, Ohm's law can be used to calculate it.

Resistance in a parallel circuit is a more complicated matter (see Figure 7.6). Three rules apply:

1. Where all resistances are equal, total resistance equals the resistance of one resistor divided by the number of resistors ($R_T = \dfrac{R_1}{R_N}$).

2. Where two resistances are not equal, the total resistance equals the product of the two resistances divided by the sum of the two resistances ($R_T = \dfrac{R_1 \times R_2}{R_1 + R_2}$).

3. Where three or more resistances are not equal, 1 divided by total resistance equals the sum of 1 divided by each resistance ($\dfrac{1}{R_T} = \dfrac{1}{R_1} + \dfrac{1}{R_2} + \ldots \dfrac{1}{R_N}$).

For a quick check, note that the total resistance in a parallel circuit will always be less than the lowest parallel resistance, but the resistance in a series circuit will always be higher than the highest. Refer to Figure 7.6 for help with the three formulas.

Power

Watt's law is $P = IE$. (Power equals current times voltage.) It holds true in DC parallel circuits as well as series circuits. To find the power in a parallel circuit or a series

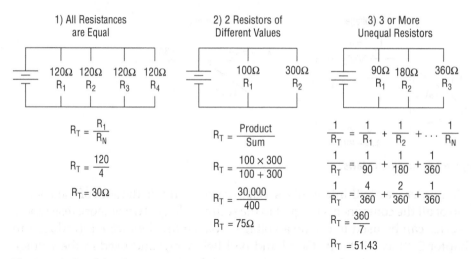

Figure 7.6: Parallel resistances

circuit, simply add the power values of each resistance together. Table 7.1 summarizes the rules for parallel and series direct current (DC) circuits.

Table 7.1: Series and Parallel Circuit Rules

VALUE/TYPE OF CIRCUIT	SERIES	PARALLEL
Current	$I_T = I_1 = I_2 = \ldots I_N$	$I_T = I_1 + I_2 + \ldots I_N$
Voltage	$E_T = E_1 + E_2 + \ldots E_N$	$E_T = E_1 = E_2 = \ldots E_N$
Resistance	$R_T = R_1 + R_2 + \ldots R_N$	1) $R_T = R_1/R_N$
		2) $R_T = $ Product/Sum
		3) $1/R_T = 1/R_1 + 1/R_2 + \ldots 1/R_N$
Power	$P_T = P_1 + P_2 + \ldots P_N$	$P_T = P_1 + P_2 + \ldots P_N$

Visit `cliffjumpertek.com/electronicFormulasTutorials` for videos and to practice calculating circuit values.

Resistance of a Conductor

The National Electrical Code (NEC) defines how much current can safely go through a wire based on its characteristics. Too much current can cause a conductor to overheat or damage equipment connected to it. Resistance (R) and *conductance* (G) of a material are inversely proportional ($\frac{1}{R} = G$). Conductance is the measure of a material's ability to transmit electrons and is measured in a unit called Siemens (S).

These are factors that affect the conductance/resistance of a conductor:

- Composition material (i.e., silver, copper, or iron)
- Length
- Cross-section area
- Temperature

For the simple projects in this book where 22-gauge hookup wire is used, the resistance of a conductor is insignificant, but in other situations it is significant. The *gauge* of a conductor identifies its size according to the American Wire Gauge (AWG) designation. AWG only applies to a solid round conductor. The sizes seem counterintuitive because a 4/0 gauge (pronounced "four aught"), which is 0000, is a much larger size than 36 gauge (4/0 = 0.46", 36 = 0.005" diameter). The larger the number, the smaller the wire. The numbers refer to the number of times that a wire had to be stretched through manufacturing dies to make it smaller.

For conductors that are not solid and round, circular mils can be used to measure size. A *circular mil* is equivalent to the area of a circle with a diameter of 0.001 inch. The area for square or any other shaped material can be mathematically converted to circular mils.

Notice in Figure 7.7 that the larger a conductor is, the less resistance it has. For the 22-gauge wire that is used in all the circuits in this book, 1,000 feet has 16.46 ohms of resistance. And 1 foot has 0.01646 ohms of resistance, and 1 inch has only 0.00137 ohms of resistance. The minute resistance of the jumpers used in these circuits is insignificant for the projects, which is why it is not considered

AWG Size:	0000	10	22	30
Size in Inches	0.460	0.1019	0.025	0.010
Resistance in Ohms for 1000 feet @ 25°C				
Copper	0.05	1.018	16.46	105.2
Aluminum	0.08	1.616	26.2	168.0

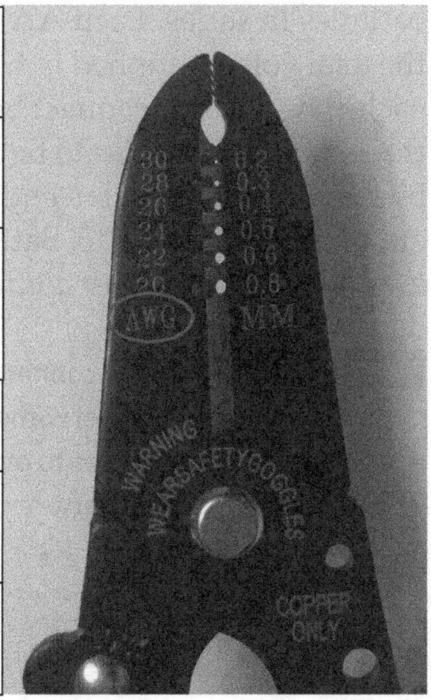

Figure 7.7: Wire gauge

in our calculations. But in a larger project where much longer wires are used, it may become important.

> **TIP**   Caution should be used when connecting heavy loads to an extension cord. Extension cords are built with certain current and resistance standards, and adding them together increases the length of the conductor, which also increases the resistance. The result is an increase in the voltage drop across the wire, which could leave too little voltage for the equipment connected to it and could cause equipment damage. Remember, a larger diameter cord (smaller AWG number) provides less resistance. OSHA.gov has more information about working safely with extension cords.

Sources in Series and Parallel

The characteristics of electricity in series and parallel have been discussed, but what happens when sources are connected in series and parallel? Whether the source is batteries or photovoltaic (PV) cells, the same rules apply. A group of batteries connected together is called a *battery bank,* and they can be configured to meet circuit needs.

Sources in Series

Suppose a circuit needs 4.5V of DC power to run. The voltage could be reduced from 9V with an appropriate resistor, but then half of the power would be wasted, dissipated as heat from the resistor. A better solution would be to connect 3 AA batteries in series. Each AA battery provides 1.5V, or 3 × 1.5V = 4.5V, meaning the total voltage applied is 4.5V. Battery holders can be purchased with leads on each end to easily connect batteries together or with premade holders for two, three, or four batteries to be inserted and automatically connected in series. To be in series, the positive end of one battery is connected to the negative end of another. See Figure 7.10. Batteries are available in many different shapes, sizes, and voltages. The most common ones are shown in Figure 7.8.

> **WARNING**   Never connect two batteries directly together (negative to positive and positive to negative) without a load. Doing so will result in a high current and could cause the batteries to overheat and ignite or explode. See Figure 7.9 for an example of improper battery connection.

Sources in Parallel

Connecting batteries in parallel means connecting all the positive ends together as one connection and the negative ends together as another connection, with a

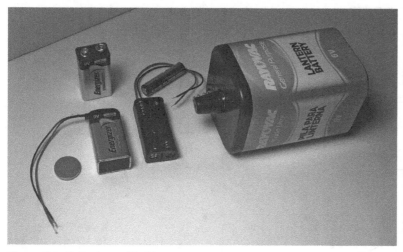

Figure 7.8: Batteries and holders

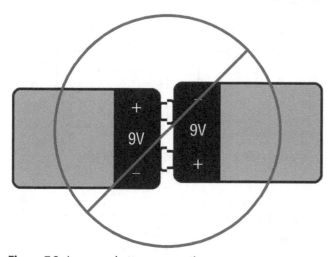

Figure 7.9: Improper battery connection

lead on each connection that is used to connect them to a circuit (see Figure 7.10). Batteries in parallel will not increase voltage, but they will increase the availability of current as well as provide more Ah (longer life) for the circuit.

Aiding and Opposing Sources

When sources are connected so that the voltage increases, they are called *series-aiding sources*. When sources are connected in such a way that the voltage is decreased, they are called *series-opposing sources*. Opposing sources negate each other to the extent of the lesser voltage. For example, if a 12V and an 8V source were connected opposing, the resultant voltage would be 4V. The direction of current flow is determined by the greater of the two opposing sources. Figure 7.11

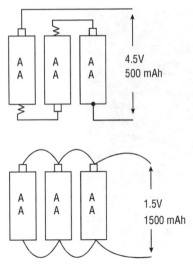

Figure 7.10: Sources in series and parallel

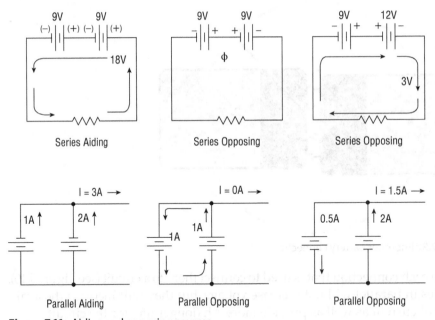

Figure 7.11: Aiding and opposing sources

illustrates aiding and opposing sources, with arrows indicating the direction of the current flow.

Parallel sources are similar, except that currents are considered rather than voltages, because voltage across a parallel circuit is equal on each branch. Equal parallel opposing sources form a closed loop; however, in unequal parallel opposing sources, the difference between the two is passed on to the connecting circuit.

Try This: Calculating Circuit Values

It's time to return to the LED circuits from the beginning of this chapter and put into action what you have learned. For many people, the chart method of solving circuit values works well. However, others like to simply write the values on a circuit drawing next to the component.

Calculating the Series Circuit

Draw a chart like the one shown in Figure 7.12 to calculate your circuit using your values. The labels across the top are the components. T stands for total, but S for source is also sometimes used. R is for resistance, whether it is a resistor or other device, although other letters may be used for other devices. Follow the process outlined here by entering your values in place of the sample values. The steps are also numbered on the chart in Figure 7.12.

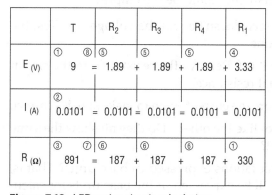

	T	R_2	R_3	R_4	R_1
$E_{(V)}$	① 9	⑧ ⑤ = 1.89	⑤ + 1.89	⑤ + 1.89	④ + 3.33
$I_{(A)}$	② 0.0101	= 0.0101	= 0.0101	= 0.0101	= 0.0101
$R_{(\Omega)}$	③ 891	⑦ ⑥ = 187	⑥ + 187	⑥ + 187	① + 330

Figure 7.12: LED series circuit calculation

1. Enter known values for the source voltage and the 330-ohm resistor, R_1.

2. Enter the measured current and extend it across because the current is equal all around a series circuit.

3. Calculate R_T using Ohm's law and the known voltage and current.

4. Calculate the voltage of R_1 using Ohm's law and the known resistance and current values.

5. LEDs are a diode, and therefore they don't truly have resistance. You'll learn about diodes in Chapter 8, "Diodes: The One-Way Street Sign," but for now we'll pretend the LEDs are replaced with an equal but unknown resistance. Calculate the three unknown resistors' voltage drop. In our example, because the source is 9V and R_1 is dissipating 3.33V, 9 – 3.333 means that there are 5.67V left for the three resistors, R_1, R_2, and R_3. Because the resistors are identical, divide 5.67 by 3 to find the expected voltage drop of each, or of course, on an actual project you could measure the

voltage drop of each and use that figure to calculate the resistance as it would be more accurate than estimating.

6. Use Ohm's law to find the resistance of each resistor, R_1, R_2, and R_3.

7. Add the resistances of the components. Do they equal the total resistance?

8. Add the voltages. Do they equal the source voltage?

Once you know what your values should be, measure the circuit. Do the voltage drops across the resistors equal what was calculated? Refer to Chapter 5, "Dim the Lights," if needed for detailed instructions on how to measure. The measured values should be close. If they are not, refer to the troubleshooting section next.

What Went Wrong

If the voltage drops that you calculated don't match your actual circuit, any of the following (and perhaps more than one) could be the culprit:

■ The stated voltage of the battery doesn't equal the voltage rise. Test the battery and use the actual value.

■ The resistor tolerance threw it off. Remove the resistor from the circuit and measure it to get its actual resistance.

■ If neither of those helps, measure the current again to make sure it's correct.

■ Possibly one of the LEDs is not like the others. Do they all show the same voltage drop when measured?

■ Were the values used in calculations expressed in base units? That is, amps, not mA? Values must all be expressed in the base units for calculating using Ohm's law.

■ Finally, recheck your math. Steps 7 and 8 of the calculation process are there as a cross-check. If they don't equal or are at least very close, suspect a math error somewhere. Occasionally small differences will occur due to rounding.

Calculating the Parallel Circuit

The parallel LED circuit in our activity is actually a complex circuit because it contains both series and parallel items. Knowing the parallel rules is necessary to determine the size of resistor to use in the circuit.

What Size Resistor Is Needed?

The source was 9V, so each branch would be 9V; however, that is too much power for an LED, so it had to be reduced. The typical red LED works at 20mA and about 2V. And, 9V less the 2V needed for the LED leaves 7V to be dissipated by a resistor. Using one resistor on the main branch is one solution. Another would be to place a resistor in each of the branches along with the LED. Both solutions would work; however, they are not equal.

The current of the branches is added to find the main branch current, so 0.02 + 0.02 + 0.02 = 0.06, or 60mA of current on the main branch. Using Ohm's law, the 7 volts divided by 0.06A gives a resistance value of 116.66 ohms repeating. Because there is some room for variance on the voltage and current used by each LED, the circuit would be fine with a 100-ohm resistor in the main branch, but a 330-ohm resistor was used for the comparison between the two circuits.

If, on the other hand, you chose to use a separate resistor on each branch, remember that each branch has 9V. The LED uses about 2V, leaving 7V to be dissipated by a resistor. So far that's the same; however, each LED needs about 20mA of current, so this time the 7V is divided by 0.02A, and the result is that a 350-ohm resistor is needed on each branch. Because there is some room for variance on the voltage and current used by each LED, a standard 330-ohm resistor would work just fine. Always refer to a component's datasheet to see how precise the calculations need to be. For this particular circuit, there was plenty of wiggle room.

Most often those are the type of questions we concern ourselves with. It's also important to ensure that the resistors used have a high enough wattage rating for the voltage and current going through them. Consider the single resistor solution. Because P = IE, the wattage rating of the single resistor would be 0.06 × 7, or 0.42 watts, so a 1/2-watt resistor would be needed. When using three resistors, however, the current is only 0.02 across each resistor, making the calculation 0.02 × 7 for a wattage of 0.14, so a 1/4-watt (0.25) resistor would be fine, but a 1/8-watt resistor would be just a bit too small.

Calculating the Complex Circuit

To solve a complex circuit (see Figure 7.13), it must be reduced to a series circuit and then calculated back to a parallel circuit. Replace the measured values in the example with your own measured values so you can practice the calculations. LEDs were used in the circuit for the visual effect, but as mentioned earlier, they are diodes and don't truly have resistance. Assume in your calculations that the LEDs on each branch are replaced with resistors of equal but unknown values. While there is more than one way to figure out the circuit values, you must always start with what is known.

1. Known: The source is 9V, and the resistor is 330 ohms. The current on the main branch was measured at 22.4mA, or 0.0224A.

2. The total resistance on the circuit can be calculated using Ohm's law, the source voltage, and the measured current (9/0.0224).

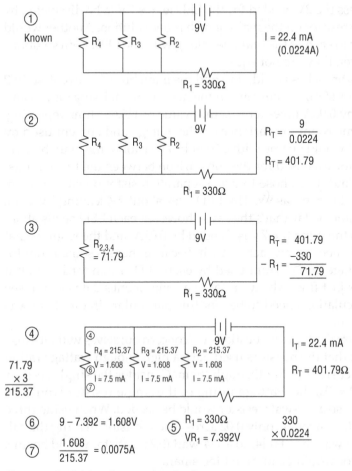

Figure 7.13: Solving the complex circuit

3. With the circuit reduced to a series circuit, the resistance of $R_{2,3,4}$ is the R_T of 401.79 ohms less 330 ohms, which equals 71.79 ohms.

4. The rule for resistance when all are equal in a parallel circuit is to divide the resistance of one by the number of resistors, so reversing that, if we multiply the resistance for $R_{2,3,4}$ by the number of resistors, the result will be the value of each resistor (71.79 * 3 = 215.37).

5. The voltage drop of the 330-ohm resistor can be calculated using the current on the main branch of 22.4mA, so R_1's voltage drop is 330 * 0.0224 = 7.392V.

6. Each branch will have the same voltage of 1.608V (9 − 7.392), because the parallel rule is that voltages on each branch are equal.

7. With the resistance and voltage of each branch, we can calculate the current of each branch, which is 0.0075A (1.608/215).

To check our circuit, the sum of the current on the branches should equal the current on the main branch, and it is close at 0.0225A. The 0.1mA difference is most likely due to rounding (0.0075A * 3).

The more you work with circuit math, the easier it becomes.

Using Common Components

In This Part

Diodes: The One-Way Street Sign

"It's never too late to start heading in the right direction."

—Seth Godin

In previous chapters, we used light-emitting diodes, but they are just one of many different types of diodes. The main purpose of a diode is not to give us different colored lights to blink and indicate things but rather to limit the flow of electrons to just one direction. Some have other uses as well. This chapter takes a closer look at the amazing diode and what it can do for our electronic circuits.

Try This: Creating a Simple Polarity Tester

In this activity, we are building a simple polarity tester. The purpose is to check the polarity of a power source, such as the repurposed power supply in Chapter 5, "Dim the Lights.". This activity works best when soldered onto a perfboard so it can be reused as needed, but it can also be created on a breadboard just to illustrate the point that diodes only allow electricity to flow in one direction.

For this project, you need the following materials:

- Small breadboard
- Green LED
- Red LED
- (2) 330-ohm resistors

- 1-½ inches of red 22-gauge wire
- 1-½ inches of blue or black 22-gauge wire
- Power source, such as a 9V or 6V battery

Optional materials include the following:

- Small perfboard
- Solder
- Soldering iron

First, create the project on a breadboard to ensure that it's working correctly (see Figure 8.1). The positions on the breadboard in Figure 8.1 are noted in the following directions.

1. Insert the LEDs so that the long lead (positive) of one LED is in the same terminal row as the short lead (negative) of the other. (The positive lead of the green LED is in position 4a, and the negative lead of the red LED is in position 4d.)

2. Insert the other lead of each LED into another terminal row but not the same terminal row. (The negative lead of the green LED is in position 10a, and the positive lead of the red LED is in position 8d.)

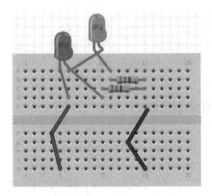

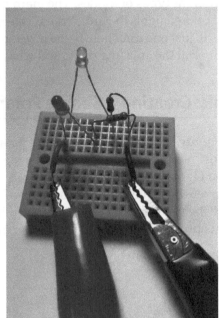

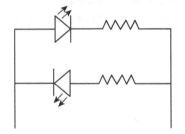

Figure 8.1: Simple polarity tester

3. Insert a current-limiting resistor in series with each LED by placing one lead of a resistor in the same terminal row as that of the second LED lead and by placing the other lead of each resistor in another terminal row together. (R_1 is inserted in 10b and 12b, and R_2 is inserted in 8c and 12c.)

4. On each of the red and black wires, strip ¼-inch off one end and ¾-inch off the opposite end.

5. Insert the ¼-inch stripped end of the red wire in the first LED lead terminal row (position 4e). Then insert the ¾-inch stripped end of the red wire in another terminal row across the dip.

6. Insert the ¼-inch stripped end of the black wire in the terminal row with the second lead of the two resistors (position 12e). Then, insert the ¾-inch-stripped end in a terminal row across the dip.

The exposed portions of the red and black wires are to provide testing points for the power source. If the positive side of a power source is touched to the red wire and the negative side of a power source is touched to the black wire, the green light illuminates to indicate that the polarity is correct—positive to red and negative to black. If the polarity is connected backward, the red light illuminates to signal that it is connected incorrectly. The voltage this circuit can handle depends on the LEDs that are used, which is explained later.

Determining Polarity Tester Maximum Voltage

The maximum and minimum voltages that can be tested with the simple polarity tester depend on the characteristics of the LEDs that are being used. The sample circuit would accept a range from approximately 3VDC to 18VDC with no problems. Remember, the circuit is quickly tested for just a few seconds and is not meant to have a permanently attached source to it as a circuit.

Although the diodes are in parallel with each other, only one conducts at a time, so each can be treated as if in a series circuit. In a series circuit, the current is the same at each component, so the current of the LED is the same as the current across the resistor. While a resistor maintains a constant resistance value, the other characteristics of the circuit can change.

Checking the datasheets of the LEDs that I used revealed that the green LED has a minimum V_F of 2.8 and a maximum V_F of 3.2. (V_F stands for voltage forward.) The LED requires the minimum voltage, or it will not light up and the current will not flow. If the maximum voltage is exceeded, the LED may work for a short period of time but will burn out either immediately or more quickly than it otherwise would have. The red LED has a minimum V_F of 2 and a maximum of 2.8. The optimum current for both LEDs is 20mA. They will most likely work at a slightly lower and slightly higher current. Some LEDs will operate with a current as low as 7mA or as high as 50mA. If in doubt, check the datasheets for the LEDs being used.

Taking both LEDs' requirements into consideration, a V_F of 2.8 would be enough to illuminate the green LED and not exceed the voltage of the red LED. Remember that those voltages are at 20mA, and there is usually some variance in current and voltage that an LED will work well with. If the source being tested were only 3V but 2.8V were needed to light the LED, then only 0.2V would be left to be dissipated by the current-limiting resistor. A voltage of 0.2V divided by 0.02A means the resistor would be only 10 ohms. Because we're using a 330-ohm resistor, if the current were 7mA, then the voltage drop of the resistor would be 2.3 volts, leaving only 0.7V for the LED. The LED may light very dimly, or not at all. A voltage of 3V or less may be too low to be tested using a 330-ohm resistor and this particular green LED.

If, on the other hand, the source were 18V, the LED would be operating closer to its maximum voltage of 3.2V, leaving 14.8V to be dissipated by the resistor. Again, using Ohm's law, the current across the resistor can be calculated and would be approximately 45mA. While some LEDs are designed to handle 50mA of current, most are not. A typical LED is more in the range of 10mA to 30mA. The LED would burn very brightly and may even exceed 3.2V; however, it would not last for an extended period of time if its maximum current were exceeded. The diode would likely break down and need to be replaced. Most LEDs will have a much higher current that they can work with for a very short period of time, perhaps 1/10 of a second, which is why this circuit is designed to touch the source only to the test points, and not connect to the source long term. Most often, an overpowered LED simply glows brighter and then goes out, or smokes and burns. On occasion, an overpowered LED has been known to explode the top off of the diode, so be careful not to overpower an LED.

Determining Resistor Needs

Remember, resistors have a wattage rating to consider. Because P = IE, in testing the 18V source for the previous circuit, the power dissipated by the resistor would be 0.67 watts (0.045A * 14.8V). So, a ¾-watt resistor would be needed if the circuit were to be connected for more than a split second.

When creating a circuit with an LED, the LED always needs a current-limiting resistor, but what value should it be? Start with the datasheet for the diode and determine the diode's typical forward voltage V_F and forward current I_F. The source voltage, V_S, also needs to be known, at which point this formula can be used:

$$R = \frac{V_S - V_F}{I_F}$$

Assume an LED with a typical V_F of 2.1V and an I_F of 20mA is supplied with a 6V source. The current-limiting resistor needed on the circuit would be 195 ohms (6 − 2.1/.02). Because 195 ohms isn't a typical resistor value, it would be fine for most purposes to use the next larger available standard size, which is most likely 200 ohms or 220 ohms. True, it provides a little extra resistance, but in a circuit with only an LED and a resistor, it would be fine. The LED would still have plenty of voltage to turn on. Remember, you must calculate the power requirement, too. For this particular situation, a ⅛-watt resistor would be fine.

Typical LED Voltages

While an LED's V_F could be anything, depending on how it was made and if it was made for a specific purpose, an ordinary, everyday LED usually falls into a voltage range based on its color. Table 8.1 lists typical voltage ranges for LEDs, based on their color. Remember, these are not the only possible voltages and currents, just typical ones.

Table 8.1: Sample LED Voltage Ranges

LED COLOR	V_F MIN @ 20MA	V_F MAX @ 20MA
Red	1.8	2.4
Yellow	1.8	2.4
Green	2.8	3.4
Blue	2.8	3.4
White	2.8	3.4

Super-bright LEDs usually have higher voltages. Although smaller in size, 3mm LEDs may have higher voltages than 5mm LEDs. Again, the best source for information is the datasheet for the particular LED being used.

Putting It on a Perfbord

If you've never used a perfboard before, think of it as being like a breadboard but without the connectors underneath. If components are inserted in adjacent holes, the connections can be easily bridged. If they are spaced two or three holes apart, a small piece of stripped wire can be used to solder them together. If farther, then a jumper wire would be needed.

Plan where your connections will be, insert the components, and then solder them together. Creating the circuit first on a breadboard may help you plan the perfboard. Simply solder on the perfboard where the terminal row connections would be on the breadboard. Solder longer jumpers where jumper wires would be.

Sometimes, components can be directly soldered together using only the component leads connecting them. In this case, I used the component leads to bridge across holes to the next component. For a circuit that will be used often, you may want to purchase a small case or use a 3D printer to make one to hold your project. Figure 8.2 shows this completed project soldered onto a perfboard. This is a fairly simple project, so it's a great one to practice using a perfboard with before attempting a more complicated perfboard project.

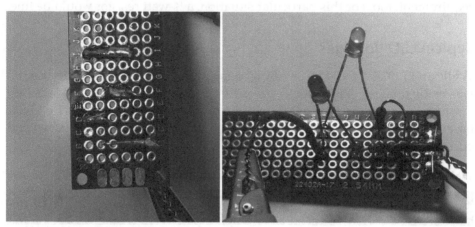

Figure 8.2: Perfboard polarity tester

LED Features

A few points about LEDs still need to be addressed. We already know that they come in different shapes, sizes, and colors. But there is more to choosing an LED.

Lumens are a measurement of the amount of light something sends out, including LEDs. Incandescent light bulbs are generally sold using watts as an indication of brightness, but as you've learned, a watt is a measure of the power converted by a device. Lumens, on the other hand, measure brightness. To give you some perspective, a 100-watt light bulb may produce 1,600 lumens of light. A projector likely is more in the area of 4,000 lumens. A compact fluorescent (CFL) light bulb uses about 26W to produce the same light as a 100W incandescent light, and an LED uses only about 22W. LED lighting is, therefore, very efficient!

What is a lumen? In general terms, a *candela* lights a sphere of a 1-foot diameter at a certain intensity, as a candle would do. A lumen is one square foot of a candela at a distance of one foot from the source. When looking at an LED or any other light source, be sure to choose the right brightness (lumens) for the job.

Unless otherwise specified, an LED's light is unidirectional. The light can be diffused by sanding an LED's surface, or diffused LEDs can be purchased. Diffused LEDs have an opaque surface, while other LEDs have a more translucent quality. Diffusing an LED gives it a greater viewing angle.

Most of the LEDs we've worked with so far can accept currents between about 10mA and 30mA. Some LEDs can accept much higher currents. Again, an LED's datasheet will list all its specifications.

LEDs can be found everywhere now. They can be purchased in strips like a rope and can be found in car headlights, household lighting, emergency lighting, and decorative lighting. In addition, they're also used for communication in computer systems. LEDs may be lighting the computer monitor that you use, and they're used in fiber-optic networks and LED printers that work similarly to laser printers. Very tiny LEDs can also be purchased on conductive stickers and used in projects like homemade greeting cards. Tiny LEDs can even be wearable when combined with conductive thread.

LEDs travel the length of the visible light spectrum and beyond. Multicolor LEDs have more than two connectors and can produce an array of colors. LEDs are also available in infrared (IR) and ultraviolet (UV) packages that are outside of the visible light spectrum and may require special handling. When using IR and UV applications, it's particularly important to be aware of the wavelength of light produced by the LED. Normally, we can see the light as color, but the longer and shorter wavelengths of IR and UV are invisible to the human eye. We can't see them, but the waves still excite molecules and are used in numerous electronics applications. IR light is what enables devices like security cameras to see in the dark. While there is a question as to whether IR LEDs can cause eye damage, at a minimum we know that IR waves produce heat, and typically they come with warnings regarding vision.

While IR wavelengths are longer than the visible light spectrum, UV light wavelengths are shorter with a higher frequency. Both UV and IR lights are already used in many devices. The best course of action when considering an IR or UV light in a project is to use due diligence and research the specifications.

In summary, when pondering electronics projects, consider the versatility of an LED and set your imagination and creativity free.

The Inner Workings of Diodes

Examine the schematic in Figure 8.1. The LEDs are connected in parallel, but in directions opposing each other. When the polarity is correct, current is able to flow through the green LED, but it is blocked by the red LED. The green LED is said to be *forward-biased*, meaning it is connected in a way that will enable electricity to flow; in this situation, it's with the positive lead of the LED toward the more positive side of the circuit. The red LED is *reverse-biased*, with its positive side toward the more negative side of the circuit. When the polarity of the source being tested isn't correct, the biasing of the two diodes is reversed, and the red diode lights to indicate such.

Diodes are made of semiconductor materials, such as silicon, germanium, and selenium, with silicon being the most popular. A silicon atom has four valence electrons that form covalent bonds with neighboring silicon atoms in a crystalline lattice structure. Because there are no free electrons or missing electrons, silicon by itself is not conductive. To be useful, silicon is doped, as is explained in the rest of this paragraph, with other elements in such a way that it will have a net positive or net negative charge. (Doping is the technical term for intentionally adding impurities to a material to change it's electrical or other properties.) Typically, boron, which has only three valence electrons, is combined with silicon, creating a net positive material. This material is called *P-type silicon* and has *holes* where electrons are missing. Other materials, such as phosphorus, which has five valence electrons, are combined with silicon to create *N-type silicon*, which has an excess of electrons. Therefore, both N-type and P-type silicon are conductive, due to either holes or excess electrons.

N-type and P-type materials are connected to form a diode. The positive end of a diode is called the *anode*, and the negative end is called the *cathode*. The area where the positive- and negative-doped materials join is called a *PN junction*. PN junctions are used in other components as well, such as bipolar junction transistors. Located at a diode's PN junction is an area called the *depletion zone*. In the depletion zone, the excess electrons are united with the holes so that there are no free electrons or holes in that area. This area is also sometimes called a *potential hill* or *potential barrier* (see Figure 8.3).

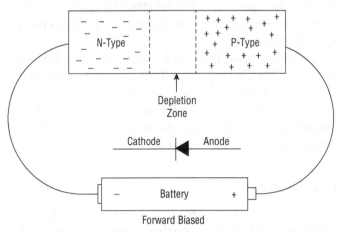

Figure 8.3: Diodes and the PN junction

When a diode is forward-biased with sufficient voltage, the holes and electrons are forced toward the center to unite, and electron flow occurs. For silicon diodes, the V_F (forward voltage) is generally around 0.6V. However, for germanium diodes, the V_F can be as little as 0.2V. This voltage is also sometimes called a *turn-on voltage*. Diodes have a maximum voltage drop, $V_{F(max)}$; a maximum forward current, $I_{F(max)}$; and a peak inverse (backward) voltage (PIV) and peak inverse current, $I_{R(max)}$. If any of the maximums are exceeded, the diode will break down. If the V_F is not met, a diode will not conduct electricity.

Determining Anode and Cathode

To be absolutely certain of the anode and cathode of a diode, the best source is the diode's datasheet. Generally speaking, however, something different will always identify the negative side of a diode. For power diodes, a stripe typically shows the cathode. For an LED, there may be a shorter lead and/or a flat spot on the side of the LED. Other packages, such as the seven-segment LED, will have either a common anode or a common cathode—the best way to be sure is to look up the datasheet for that part number. Figure 8.4 shows the anode and cathode of some common diodes.

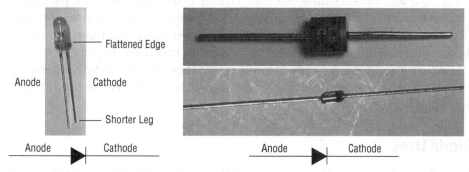

Figure 8.4: An anode and a cathode

The polarity of a diode can sometimes be determined using a multimeter. Considering that most diodes are designed to be forward-biased, measuring the resistance of a diode with the red probe on the anode and the black probe on the cathode would yield a relatively low resistance, while testing with the red probe on the cathode and the black probe on the anode would yield a much higher or infinite resistance. The resistance in forward bias may still be very high, but it is the relationship between the two readings that is important. For example, testing an LED with the red probe on the anode may yield a resistance of 1.9M ohms, while with the black lead on the anode the resistance would be infinite. This assumes, of course, that the black lead is in the COM port.

WARNING   To avoid damaging your meter, remember that before taking any resistance measurement, make sure the power to the circuit is off.

Some meters have specific settings or places to insert the component for testing devices like diodes and transistors. Check the meter's documentation to know for sure.

Types of Diodes

Various types of diodes serve different purposes and have different characteristics depending on the material they are constructed of and how they are constructed. Germanium diodes are often used for radio frequency signals because of their low V_F. Diodes known as Schottky diodes are chosen for faster switching speeds. And, rectifier diodes handle higher voltage and current. Some diodes, such as Zener diodes, are designed to be connected reverse-biased. Photodiodes conduct current based on the light that strikes their surface. IR diodes and laser diodes can be used as alarm system switches, among other things.

The best way to determine whether a diode is the right one for a particular application is to check the specific diode's datasheet. The datasheet contains all the minimum and maximum information listed above and more. Figure 8.5 shows several different diodes, but keep in mind that there are others.

Diodes have different symbols to represent them in schematics depending on the type of diode required. Figure 8.6 shows the symbols used for different types of diodes.

Diode Uses

Zener diodes in series with a resistor are often used as voltage regulators, where a constant voltage is needed by a load (see Figure 8.7). The circuit load is placed in parallel to the Zener diode. In a parallel circuit, the parallel branches have equal voltages, so whatever voltage the Zener diode has, the load will also have. If the source voltage changes, the Zener maintains a constant voltage. If the circuit current changes, the current across the Zener diode will change, keeping the voltage at the load constant.

Power diodes in series can be used to produce consistent voltage drops in a circuit. This can act as a voltage regulator if they are in parallel with another part of a circuit or a load, similar to the Zener diode (see Figure 8.8).

Because diodes allow current to flow in only one direction, they can be used to rectify an AC signal to a DC signal. A bridge rectifier combines four diodes to convert the negative side of a sine wave to positive (see Figure 8.9). This is called *full wave rectification*. Capacitors are used to filter the resulting wave into a smoother signal.

Properly placed diodes on the output side of a transformer can double or triple voltage output and are used in power supplies. Transformers are discussed in Chapter 16, "Transformers and Power Distribution."

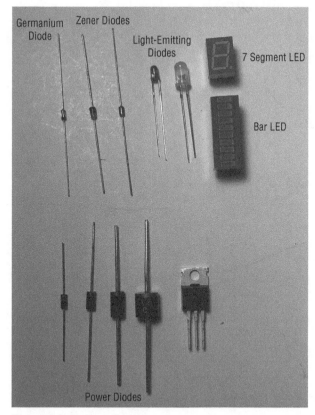

Figure 8.5: Various diodes

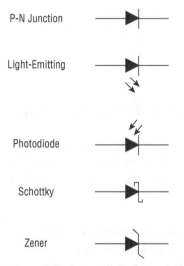

Figure 8.6: Common diode symbols

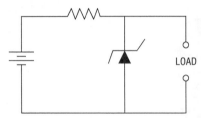

Figure 8.7: A Zener voltage regulator

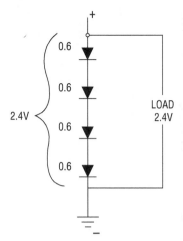

Figure 8.8: A power diode voltage divider/regulator

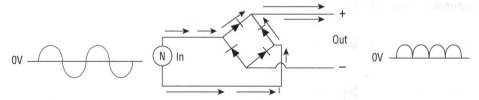

Figure 8.9: A bridge rectifier

In radio systems, diodes are used on the receiver end to restore the output back to its original signal.

Laser diodes are used in many applications, from optical computer equipment to scientific instruments used for measurement to levels for construction sites and even precision cutting applications. Laser beams can also be used in security systems and surgery.

WARNING   Lasers can be dangerous and can be powerful! Some are so strong that they're used to cut metal. Laser light can cause immediate and permanent eye damage. Special eye protection must be worn at all times when working with dangerous classes of lasers.

One source of information about laser safety is the Occupational Safety and Health Administration (OSHA) website (www.osha.gov/SLTC/laserhazards/). Laws exist in the United States and other countries regarding laser use, so if you're planning an application involving lasers, make sure that it's both safe and legal.

A flyback diode is a diode placed in parallel with a motor or other large inductor to avoid damage to the rest of the circuit. Motors and inductors create large magnetic fields while they are being used, but when the power is turned off, the magnetic field collapses, and a sudden voltage spike occurs in the circuit. A flyback diode is placed in opposite polarity to the source, so when the circuit is turned off, the diode blocks the voltage spike from damaging the rest of the circuit. Figure 8.10 shows a schematic using a flyback diode (D_1). L_1 is an inductor, which can be the coil of a motor or other device as well as a discrete inductor.

Flyback diodes are also used to reduce electrical noise in circuits with devices like relays. Relays are discussed in more detail in Chapter 11, "The Magic of Magnetism."

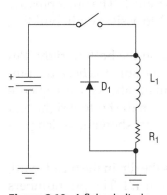

Figure 8.10: A flyback diode

Try This: Using a Seven-Segment LED

Seven-segment LEDs and bar LEDs are groups of individual LEDs in a package that share either a common anode or a common cathode. Once you've mastered using them, chances are you'll think of a myriad of ways they can be used. They are particularly fun to work with using the Arduino. Both seven-segment and bar LEDs can be used in this activity.

For this project, you need the following materials:

- Breadboard
- Arduino Uno, PC with IDE and USB cable
- Seven-segment LED (LTS-4301JF)
- (8) 100-ohm resistors

Optional materials include the following:

- Bar LED (LTA-1000HR)
- (8) 100-ohm resistors

Seven-Segment LED

If the seven-segment LED that you're using has a different part number, research that part number to find the datasheet and adjust the directions as necessary.

To calculate the resistor value, take the source voltage, in this case 5V from the Arduino, and subtract the typical voltage of the LED, 2.01V. Divide the result by the desired current of 25mA, for a resistance value of 119 ohms. Because 119 isn't a standard number, using a 100-ohm resistor for this application is fine. (There are occasions when the resistor value must be very precise, but this isn't one.) Substitute values for the LED you have if it's a different one.

The LTS-4301JF LED has common cathodes at pins 8 and 3. The other pins are the anode connections for each LED. Use Figure 8.11 to assist you in building this project.

Orient the LED so that the decimal point (DP) is on the bottom right. Pin number 1 is on the top left. (If using a different LED, be sure to check the pin-out on the datasheet for that LED.) From there, the pin numbers go in sequence counterclockwise, so pin 5 is on the bottom left, pin 6 on the bottom right, and pin 10 on the top right. Refer to the datasheet for a chart showing which pins are attached to each section of the LED.

1. Begin by placing the seven-segment LED across the dip in the breadboard, with pin 1 at location 12e and pin 10 at location 12f. The pin numbers reference the demonstration breadboard numbers. Yours may be different, so remember to adjust the connections as necessary for your specific part number.

2. Connect a black jumper wire from pins 3 and 8 (which are in row 14 on opposite sides of the breadboard) to the negative rail on each respective side of the breadboard. Jumper across the bottom of the breadboard to connect the two negative rails.
   Hint: It may be helpful to shorten the leads of the resistors so they won't short out against each other or be accidentally pulled out of their connections. They'll also look tidier.

3. Place a 100-ohm resistor between each of the following locations:

12b to 8b	13g to 9g
13d to 9d	12g to 8g
15d to 19d	15g to 19g
16b to 20b	16i to 20i

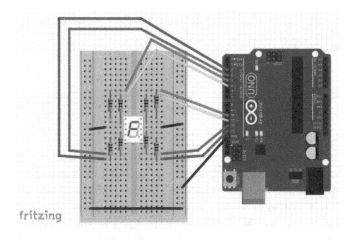

Figure 8.11: Seven-segment LED wiring

4. Connect a jumper between an Arduino ground pin and a ground rail on the breadboard.

5. Make the following connections between the digital I/O pins on the Arduino board and the breadboard:

Arduino	to	Breadboard
2		20a
3		19a
4		9e
5		8e

10	8j
11	9j
12	19j
13	20j

The connections should look like Figure 8.11.

Now for the programming. In this sketch, the LED simply counts each time it loops through the code. However, think about the possibilities. It could be rearranged to show a temperature being read from other analog in pins and display it on multiple LEDs. It could be used to display a count of the number of times a button is pushed or the number of times that a laser beam is broken as someone walks through a doorway. The possibilities are nearly endless and are limited only by your imagination.

Open the Arduino IDE and enter the following sketch:

```
// First, link the segments to a pin.
// The letters are the names of segments on the LED.
// Each segment is linked to a pin on the LED (see the datasheet)
// Each of the LED's pins are connected to a pin on the Arduino
// The List below links the LED segment to an Arduino I/O pin.

int A=10;
int B=11;
int C=12;
int D=3;
int E=2;
int F=5;
int G=4;
int DP=13;


void setup() {
  // Set the mode of the I/O pins to output. Remember you can copy/
paste.

  pinMode(A, OUTPUT);
  pinMode(B, OUTPUT);
  pinMode(C, OUTPUT);
  pinMode(D, OUTPUT);
  pinMode(E, OUTPUT);
  pinMode(F, OUTPUT);
  pinMode(G, OUTPUT);
  pinMode(DP, OUTPUT);

}
```

```
void loop() {
  // The LED will count up from 0 to 9 then flash and start over again.
  // It may be helpful for you to write down what segments equal each
number 0 through 9


// making sure all are off
  digitalWrite(DP, LOW);
  digitalWrite(A, LOW);
  digitalWrite(B, LOW);
  digitalWrite(C, LOW);
  digitalWrite(D, LOW);
  digitalWrite(E, LOW);
  digitalWrite(F, LOW);
  digitalWrite(G, LOW);
  delay(1500);

// flashing ON
  digitalWrite(DP, HIGH);
  digitalWrite(A, HIGH);
  digitalWrite(B, HIGH);
  digitalWrite(C, HIGH);
  digitalWrite(D, HIGH);
  digitalWrite(E, HIGH);
  digitalWrite(F, HIGH);
  digitalWrite(G, LOW);
  delay(500);

// flashing OFF
  digitalWrite(A, LOW);
  digitalWrite(B, LOW);
  digitalWrite(C, LOW);
  digitalWrite(D, LOW);
  digitalWrite(E, LOW);
  digitalWrite(F, LOW);
  digitalWrite(G, LOW);
  delay(500);

// 0
  digitalWrite(A, HIGH);
  digitalWrite(B, HIGH);
  digitalWrite(C, HIGH);
  digitalWrite(D, HIGH);
  digitalWrite(E, HIGH);
  digitalWrite(F, HIGH);
  digitalWrite(G, LOW);
  delay(1000);

  // 1
```

```
  digitalWrite(A, LOW);
  digitalWrite(B, HIGH);
  digitalWrite(C, HIGH);
  digitalWrite(D, LOW);
  digitalWrite(E, LOW);
  digitalWrite(F, LOW);
  digitalWrite(G, LOW);
  delay(1000);

// 2
  digitalWrite(A, HIGH);
  digitalWrite(B, HIGH);
  digitalWrite(C, LOW);
  digitalWrite(D, HIGH);
  digitalWrite(E, HIGH);
  digitalWrite(F, LOW);
  digitalWrite(G, HIGH);
  delay(1000);

// 3
  digitalWrite(A, HIGH);
  digitalWrite(B, HIGH);
  digitalWrite(C, HIGH);
  digitalWrite(E, HIGH);
  digitalWrite(D, LOW);
  digitalWrite(F, LOW);
  digitalWrite(G, HIGH);
  delay(1000);

// 4
  digitalWrite(A, LOW);
  digitalWrite(B, HIGH);
  digitalWrite(C, HIGH);
  digitalWrite(D, LOW);
  digitalWrite(E, LOW);
  digitalWrite(F, HIGH);
  digitalWrite(G, HIGH);
  delay(1000);

// 5
  digitalWrite(A, HIGH);
  digitalWrite(B, LOW);
  digitalWrite(C, HIGH);
  digitalWrite(E, HIGH);
  digitalWrite(D, LOW);
  digitalWrite(F, HIGH);
  digitalWrite(G, HIGH);
  delay(1000);

// 6
```

```
    digitalWrite(A, HIGH);
    digitalWrite(B, LOW);
    digitalWrite(C, HIGH);
    digitalWrite(D, HIGH);
    digitalWrite(E, HIGH);
    digitalWrite(F, HIGH);
    digitalWrite(G, HIGH);
    delay(1000);

  // 7
    digitalWrite(A, HIGH);
    digitalWrite(B, HIGH);
    digitalWrite(C, HIGH);
    digitalWrite(D, LOW);
    digitalWrite(E, LOW);
    digitalWrite(F, LOW);
    digitalWrite(G, LOW);
    delay(1000);

  // 8
    digitalWrite(A, HIGH);
    digitalWrite(B, HIGH);
    digitalWrite(C, HIGH);
    digitalWrite(D, HIGH);
    digitalWrite(E, HIGH);
    digitalWrite(F, HIGH);
    digitalWrite(G, HIGH);
    delay(1000);

  // 9
    digitalWrite(A, HIGH);
    digitalWrite(B, HIGH);
    digitalWrite(C, HIGH);
    digitalWrite(D, LOW);
    digitalWrite(E, LOW);
    digitalWrite(F, HIGH);
    digitalWrite(G, HIGH);
    delay(1000);

}
```

Compile and upload the sketch. The 0 should flash and then count up from 0 through 9 and start over.

Bar LED

Now for the bar LED (see Figure 8.12). The bar LED is far simpler to work with. Each LED has a separate anode and cathode like any other LED. To determine pin 1, look for something different on the LED package. For this package, that

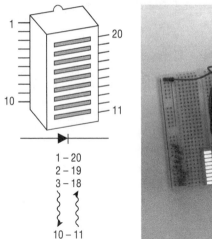

Figure 8.12: A bar LED

marking is an extra piece of plastic at the top, similar to the divot in an IC. Once it's located, it's easy to figure out that pin 1 is to its left. From that point, the pin numbers increment counterclockwise just like an IC.

1. Place the bar LED anywhere across the dip in the breadboard. For the demo, pin 1 is in position 1g, and pin 20 is in pin 1f.

2. For this project, jumper all the cathodes (pins 11 through 20) to the negative rail using a 100-ohm resistor for each, and jumper the negative rail to the ground pin on the Arduino.

3. Starting with digital I/O pin 1, jumper pin 1 on the Arduino to pin 1 on the LED and continue up to pin 10.

Open the IDE and begin a new sketch. The sketch (i.e., code) will light one LED at a time, keeping each LED on until all are lit. Then, they will flash, go out, and start over.

Enter the code as follows:

```
// Initializing the pins isn't necessary since pin 1 on the Arduino =
pin 1 on the LED.

void setup() {
  // set the pin modes to output.

pinMode(1, OUTPUT);
pinMode(2, OUTPUT);
pinMode(3, OUTPUT);
pinMode(4, OUTPUT);
pinMode(5, OUTPUT);
pinMode(6, OUTPUT);
```

```
  pinMode(7, OUTPUT);
  pinMode(8, OUTPUT);
  pinMode(9, OUTPUT);
  pinMode(10, OUTPUT);

}

void loop() {
  // put your main code here, to run repeatedly:

  // making sure all are off
  digitalWrite(1, LOW);
  digitalWrite(2, LOW);
  digitalWrite(3, LOW);
  digitalWrite(4, LOW);
  digitalWrite(5, LOW);
  digitalWrite(6, LOW);
  digitalWrite(7, LOW);
  digitalWrite(8, LOW);
  digitalWrite(9, LOW);
  digitalWrite(10, LOW);
  delay(1500);

// flashing ON
  digitalWrite(1, HIGH);
  digitalWrite(2, HIGH);
  digitalWrite(3, HIGH);
  digitalWrite(4, HIGH);
  digitalWrite(5, HIGH);
  digitalWrite(6, HIGH);
  digitalWrite(7, HIGH);
  digitalWrite(8, HIGH);
  digitalWrite(9, HIGH);
  digitalWrite(10, HIGH);
  delay(1500);

  // making sure all are off
  digitalWrite(1, LOW);
  digitalWrite(2, LOW);
  digitalWrite(3, LOW);
  digitalWrite(4, LOW);
  digitalWrite(5, LOW);
  digitalWrite(6, LOW);
  digitalWrite(7, LOW);
  digitalWrite(8, LOW);
  digitalWrite(9, LOW);
  digitalWrite(10, LOW);
  delay(500);

    // 1
```

```
digitalWrite(1, HIGH);
digitalWrite(2, LOW);
digitalWrite(3, LOW);
digitalWrite(4, LOW);
digitalWrite(5, LOW);
digitalWrite(6, LOW);
digitalWrite(7, LOW);
digitalWrite(8, LOW);
digitalWrite(9, LOW);
digitalWrite(10, LOW);
delay(500);

  // 2
digitalWrite(1, HIGH);
digitalWrite(2, HIGH);
digitalWrite(3, LOW);
digitalWrite(4, LOW);
digitalWrite(5, LOW);
digitalWrite(6, LOW);
digitalWrite(7, LOW);
digitalWrite(8, LOW);
digitalWrite(9, LOW);
digitalWrite(10, LOW);
delay(500);

  // 3
digitalWrite(1, HIGH);
digitalWrite(2, HIGH);
digitalWrite(3, HIGH);
digitalWrite(4, LOW);
digitalWrite(5, LOW);
digitalWrite(6, LOW);
digitalWrite(7, LOW);
digitalWrite(8, LOW);
digitalWrite(9, LOW);
digitalWrite(10, LOW);
delay(500);

      // 4
digitalWrite(1, HIGH);
digitalWrite(2, HIGH);
digitalWrite(3, HIGH);
digitalWrite(4, HIGH);
digitalWrite(5, LOW);
digitalWrite(6, LOW);
digitalWrite(7, LOW);
digitalWrite(8, LOW);
digitalWrite(9, LOW);
digitalWrite(10, LOW);
delay(500);
```

```
    // 5
digitalWrite(1, HIGH);
digitalWrite(2, HIGH);
digitalWrite(3, HIGH);
digitalWrite(4, HIGH);
digitalWrite(5, HIGH);
digitalWrite(6, LOW);
digitalWrite(7, LOW);
digitalWrite(8, LOW);
digitalWrite(9, LOW);
digitalWrite(10, LOW);
delay(500);

    // 6
digitalWrite(1, HIGH);
digitalWrite(2, HIGH);
digitalWrite(3, HIGH);
digitalWrite(4, HIGH);
digitalWrite(5, HIGH);
digitalWrite(6, HIGH);
digitalWrite(7, LOW);
digitalWrite(8, LOW);
digitalWrite(9, LOW);
digitalWrite(10, LOW);
delay(500);

    // 7
digitalWrite(1, HIGH);
digitalWrite(2, HIGH);
digitalWrite(3, HIGH);
digitalWrite(4, HIGH);
digitalWrite(5, HIGH);
digitalWrite(6, HIGH);
digitalWrite(7, HIGH);
digitalWrite(8, LOW);
digitalWrite(9, LOW);
digitalWrite(10, LOW);
delay(500);

    // 8
digitalWrite(1, HIGH);
digitalWrite(2, HIGH);
digitalWrite(3, HIGH);
digitalWrite(4, HIGH);
digitalWrite(5, HIGH);
digitalWrite(6, HIGH);
digitalWrite(7, HIGH);
digitalWrite(8, HIGH);
digitalWrite(9, LOW);
digitalWrite(10, LOW);
delay(500);
```

```
        // 9
digitalWrite(1, HIGH);
digitalWrite(2, HIGH);
digitalWrite(3, HIGH);
digitalWrite(4, HIGH);
digitalWrite(5, HIGH);
digitalWrite(6, HIGH);
digitalWrite(7, HIGH);
digitalWrite(8, HIGH);
digitalWrite(9, HIGH);
digitalWrite(10, LOW);
delay(500);

        // 10
digitalWrite(1, HIGH);
digitalWrite(2, HIGH);
digitalWrite(3, HIGH);
digitalWrite(4, HIGH);
digitalWrite(5, HIGH);
digitalWrite(6, HIGH);
digitalWrite(7, HIGH);
digitalWrite(8, HIGH);
digitalWrite(9, HIGH);
digitalWrite(10, HIGH);
delay(500);

}
```

When combined with the Arduino's analog input pins and a bit of extra programming, the bar could indicate almost anything measurable. Get inventive! Have fun with it and make it your own!

Transistors

"How do you define real? If you're talking about what you can feel, what you can smell, can taste and see, then real is simply electrical signals interpreted by your brain."

—Morpheus, *The Matrix*

Most likely you use transistors every day, unaware. Transistors are everywhere in our devices. Literally billions of microscopic transistors are part of the processors that are the brains of our computers, cell phones, and other smart devices that we use daily.

Transistors have two main jobs. They can act as switches or can be used to amplify a signal. This chapter takes a brief look at both uses.

Try This: Using a Transistor as an Amplifier

This classic project for electronics beginners showcases a transistor as an amplifier and is a lot of fun. For this project, you will need the following materials:

- Breadboard
- BC547 transistor
- 9V battery
- 100-ohm resistor
- 330-ohm resistor
- 10k-ohm resistor

- Light-emitting diode (LED), any color
- Assorted jumper wires

First, to make sure that the transistor is the correct one, look at the numbers on the flat side of the transistor (see Figure 9.1). The BC547 shown is in a package (shape and size) known as a TO-92. A quick look at its datasheet shows the pinout for the transistor. Looking at the flat side of the transistor, the legs from left to right are named *collector*, *base*, and *emitter*. Other transistors, such as the 2N3904, have legs in the opposite order—left to right: emitter, base, and collector—so it's always a good idea to check the datasheet for the transistor being used. On a circuit diagram, the emitter is always the leg with the arrow. The base is in the middle, and the collector is the other leg (see Figure 9.2).

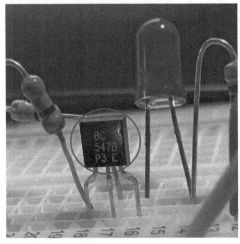

Figure 9.1: A transistor face

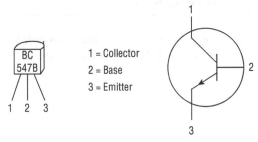

1 = Collector
2 = Base
3 = Emitter

Figure 9.2: A transistor pinout

The letter after the 547 designates the variations in the transistor. For example, the BC547A has a gain of up to 220, while the BC547B has a gain between 200 and 450. The BC547C can have a gain up to 800. This information is found on the datasheet for the transistor. But what is gain?

Gain is represented by the symbol beta β and is a factor that represents the difference between the current applied to the base of the transistor and the current that flows between the collector and emitter.

$$Current\ Gain = \frac{Output\ Current}{Input\ Current}$$

For example, if the current applied to the base is 1mA and the current measured coming from the collector is 200mA, then the transistor has a gain of 200 ($\frac{0.200A}{0.001A}$). Note, gain doesn't have a unit; it is simply a factor.

Follow these steps to build the circuit. Refer to Figure 9.3 to aid in placing the components. The connections referenced are for the sample board, so adjust the connections to match your board as necessary.

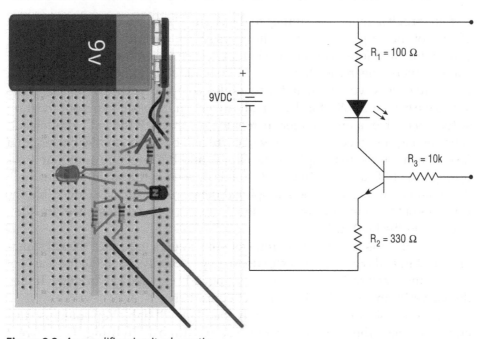

Figure 9.3: An amplifier circuit schematic

1. Place a small red jumper between the positive power rail and 12j.

2. Place the 100-ohm resistor between 12h and 14h.

3. Place the long lead (i.e., positive side) of the LED in 14g and the short lead in 16g.

4. Place the transistor so that the collector is in 16j (i.e., the same row as the LED's negative lead), the base is in 17j, and the emitter is in 18j.

5. Place the 10k-ohm resistor between 17g (i.e., the same row as the transistor base) and 23g.

6. Place the 330-ohm resistor between 18g (i.e., the same row as the transistor emitter) and 20g.

7. Place a small black jumper between 20j and the negative rail.

8. Strip a quarter inch off of both ends of a red jumper and a black jumper, each about 3 inches long.

9. Place one end of the red jumper in the power rail.

10. Place one end of the black jumper in 23f (i.e., the same row as the 10k resistor).

11. Insert the leads from the 9V battery into the positive and negative rails.

Your circuit should look similar to the circuit shown in Figure 9.4. One end of each of the long red and black jumpers should be sticking out freely. Use your thumb and forefinger of each hand to pinch the free ends of the red and black jumpers, one in each hand. The LED should light faintly. A small current is running across your skin and through R3, completing the circuit that flows across the base and collector. This small current on the base "turns on" the transistor and enables a much larger current to flow through the LED, emitter, and collector. Figure 9.5 depicts the flow of current. Notice the smaller current that flows when your skin completes the circuit, and the larger current that flows through the LED as the path from C to E is opened by the small current on B. A similar circuit in a classroom amplified the signal to such an extent that the free ends of the jumpers were connected through a string of 21 students holding hands, and the LED still lit.

If the LED fails to light or is dim, wet your fingers to reduce the resistance of your skin and try again. If it still doesn't light, briefly touch the free leads of the red and black resistors together, and the LED should light brightly. If they do not, verify that the circuit is connected properly.

The more resistance, the more faintly the LED will light. You could use this circuit to check the relative resistance across almost anything. The current across your skin is extremely small, yet enough current is applied to the base of the transistor to allow electrons to flow from the emitter to the collector and faintly light the LED.

Measuring the demo circuit revealed a current of 5μA (0.000005 A) at the base and a current across the LED of 2mA (0.002A), providing a gain of 400 $(\frac{0.002A}{0.000005A})$.

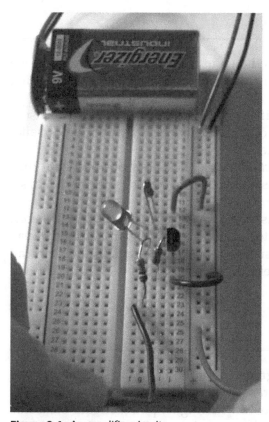

Figure 9.4: An amplifier circuit

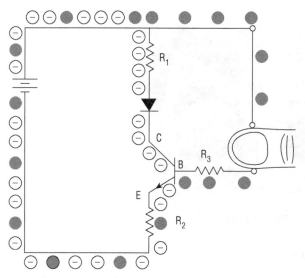

Figure 9.5: An amplifier current flow

The Purpose of Transistors

Many different transistors exist, as shown in the next section, but why? One reason is that different characteristics of electricity come into play, depending on the type of circuit.

The transistor used in the amplifier circuit is a bipolar junction transistor. Bipolar junction transistors (BJTs) work based on current. They enable a smaller current to control a much larger one.

Other transistors called *field effect transistors* (FETs) rely on voltage to "turn on," and like a BJT, they can use a small voltage to control a circuit that has a much larger voltage.

When used as an amplifier, transistors provide a safety factor for someone working with the device the transistor is attached to. They can also take a weak signal and make it much stronger, such as a weak radio signal. Transistors can also work as simple switches in an electronic circuit and be used to create complex digital logic circuits.

Types of Transistors

Visually, transistors are easily segregated into signal transistors and power transistors. See Figure 9.6. Signal transistors are the type used for low-power applications. Power transistors are used when higher-power circuits come into play. Transistors come in various packages, such as the TO-92, TO-18, and TO3, just to name a few.

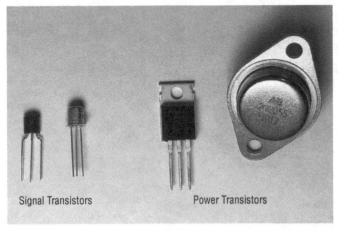

Signal Transistors   Power Transistors

Figure 9.6: Signal and power transistors

Distinguishing Transistor Types

Transistors can also be divided by type. As previously mentioned, there are BJTs and FETs. FETs can further be divided into metal-oxide semiconductor FETs (MOSFETs) and junction FETs (JFETs). MOSFETs also come in depletion mode and enhancement mode, and there are power MOSFETs and insulated gate BJTs. All of these transistor types have positive and negative varieties. For BJT varieties, the negative and positive types are NPN and PNP. For FET varieties, the negative and positive types are called *N-channel* and *P-channel*. See Figure 9.7. The best way to figure out what you have is to research the part number online. Most transistors have a part number stamped somewhere on the package.

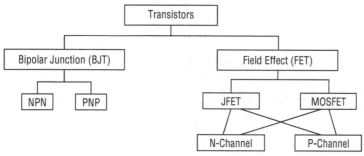

Figure 9.7: The hierarchy of transistors

Determining Transistor Connections

As mentioned, the BJT connectors are emitter, base, and collector. The connectors on a FET are called *gate*, *drain*, and *source*.

To determine which lead is which, locate the datasheet for that particular transistor. Somewhere on the datasheet should be a pinout showing which physical pin equals the symbolic pin.

Once the pin is known, it must be compared to the schematic to properly connect it. Figure 9.8 shows several common schematic symbols used to identify transistors.

Bipolar Junction Transistors

First, look at the BJT symbols. The arrow is always the emitter leg. If the arrow points in, the transistor is PNP. If the arrow points out, it is NPN. A trick to help remember this is "NPN Never Points iN."

Transistor Symbols

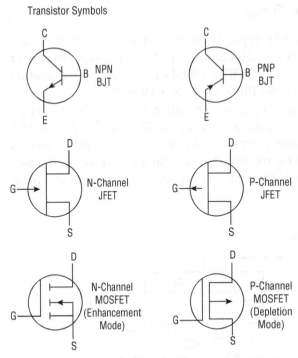

Figure 9.8: Transistor symbols

BJTs are made of doped silicon similar to the diodes discussed in Chapter 8, "Diodes: The One-Way Street Sign." The base of an NPN transistor is made of positively doped material sandwiched between two pieces of negatively doped material. It acts similarly to an open switch and does not allow current to flow from the emitter to the collector unless a positive voltage of at least 0.6V is applied to the base. Generally speaking, the greater the voltage at the base, the greater the number of electrons that are able to flow between the emitter and collector until the transistor's maximum is reached, which is known as *saturation*. Additionally, the voltage at the collector must be more positive and greater than the voltage at the emitter.

The base of a PNP transistor is made of negatively doped material. A PNP transistor acts similarly to a closed switch, allowing current to flow from emitter to collector *until* a current is applied at the base. When current is applied at the base, the current able to flow through the emitter-collector path is reduced, as current to the base is increased. The voltage applied to the base of the PNP transistor must be negative in relationship to the voltage at the emitter.

Field Effect Transistors

Now, let's look again at Figure 9.8, this time focusing on the symbols for FETs. The easiest distinction between a JFET and a MOSFET is the different gate representation. Also, notice that with FETs, the N versions have arrows that point in, while the P versions have arrows that point out, which is opposite of the emitter arrows in the BJTs. The bottom two symbols are the enhancement mode and depletion mode symbols. While there are positive and negative versions of both, only one of each is shown to differentiate between them. Notice the three lines to the right of the gate on the enhancement mode MOSFET, while there is a solid line on the depletion mode MOSFET.

Again, to determine which pin is which on any particular device, research online to find a datasheet for the device. Digikey.com and mouser.com are great places to start, or you can use your favorite browser. Once you find the part number, click the link to the datasheet.

While BJTs rely on current to control a circuit, FETs rely on voltage. Like BJT transistors, N-channel and P-channel FETs work in opposite ways. With an N-channel MOSFET, current will not flow until a voltage is applied to the gate. The typical voltage is small, such as 5V. The source pin is connected to circuit ground. The drain pin connects to the negative side of the load, and the positive side of the load connects to the positive side of the load's power supply. Although there is a complete circuit from the positive side of the power supply through the load and the transistor to the negative side of the circuit, electrons can't flow through the transistor until a voltage is applied to the gate.

When no voltage is applied to the gate, the conductivity between the drain and source is very low. When voltage *is* applied to the gate and current flows from the gate to the source, the conductance of the transistor increases rapidly and effectively opens the path from the drain to the source, enabling the connected load to function. Figure 9.9 depicts the flow of electrons when a voltage is applied to the gate of an N-channel MOSFET. It shows the drain to source voltage at 12V, but this could be hundreds of volts depending on the capabilities of the transistor being used.

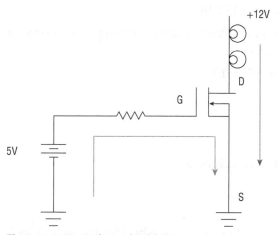

Figure 9.9: An N-channel MOSFET

While the gate voltage is typically around 5V, the load side of the transistor varies greatly. An example of a high-power transistor is the Siemens Corporation BUZ78, which can handle voltages as high as 800V on the load side (although the maximum gate voltage is only 20V). Another MOSFET, the TP0604N3 from Microchip Technology, has a maximum drain source voltage of -40V, while the maximum gate voltage is still +/- 20V. When the load is an AC load, it may be better to use a transistor to switch a relay on and off and have the relay connected to the AC load. (Relays will be covered in Chapter 11, "The Magic of Magnetism.")

The signal (gate) side of the MOSFET (or any other transistor) could be another circuit, such as a 555 timer circuit, or even output from a computer or Arduino board. It could be almost anything, even a pressure switch that someone steps on when they enter a doorway, or a circuit that reacts when a laser beam or magnetic field is disrupted. Transistor uses are limited only by a user's imagination, so start getting creative!

Try This: Using a Transistor as a Switch

This activity is designed to help you become more comfortable with power MOSFETs. It's a simple circuit that uses a 6V source on the gate of an N-channel power MOSFET to control a string of incandescent lights running on a higher voltage. The lights in the demo circuit are running on a 12V 300mA source that was the repurposed power supply in Chapter 5, "Dim the Lights." A different load and corresponding power source could be used as long as they are within the limits of the MOSFET and the load circuit is properly designed.

For this project, you need the following materials:

- Breadboard with two sets of power rails
- (2) 6V incandescent lights, or one 12V light
- An appropriate 12V DC power source (could be a repurposed source or two 6V lantern batteries in series)
- P30N06LE N-channel power MOSFET
- 6V lantern battery
- Small slider switch
- 10k-ohm resistor
- Assorted jumper wires and wire strippers

Verifying the Data

First, consider the MOSFET being used. Locate the datasheet for it online. A look at the datasheet for the P30N06LE MOSFET indicates the following:

- A power MOSFET
- N-channel
- Enhancement mode—off until a voltage is applied to the gate
- A logic level gate, so the voltage gate to source threshold (VGSTH) is less than 5V, meaning the transistor can be "turned on" by devices such as an Arduino or Raspberry Pi that work on small voltages
- Voltage maximum of 60V, drain to source
- Current maximum of 30A, drain to source
- TO-220AB (i.e., the package)
- A resistance between the drain and the source when the transistor is on ($R_{DS}ON$) of 47 mohms

The resistance between the drain and the source when the transistor is on ($R_{DS}ON$) can be found on the datasheet. For this transistor, the $R_{DS}ON$ is 0.047 ohms. Why does it matter? Transistors can get hot and sometimes need a heat sink to help dissipate the heat so the transistor is not destroyed. The $R_{DS}ON$ is used to determine how hot the transistor will get. A typical MOSFET has a metal drain flange at the top. The drain flange has a hole in it for attaching the MOSFET to a metal heat sink with a screw, as shown in Figure 9.10. Some heat sinks have multiple metal "fingers" that increase the surface area to more effectively dissipate heat, which is the purpose of any heat sink.

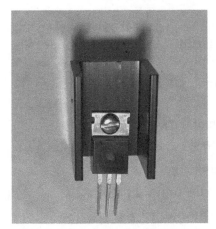

Figure 9.10: A transistor and heat sink

First, calculate the heat that will be created at the transistor. The formula for power (heat) is taken from the power wheel and is $P = I^2R$. For a transistor, R is equal to the transistor's $R_{DS}ON$, and I is equal to the current the load will draw. This circuit's load won't draw any more than 300mA, so $P = 0.3^2 * 0.047$;

hence, the total power for this circuit that must be dissipated by the transistor is 0.00423, or 4.23mW of power.

To determine how much power the transistor is capable of handling, look again at the datasheet. Locate the maximum operating (junction) temperature and the junction-to-ambient coefficient maximum. To find the most heat the transistor can handle without a heat sink, subtract the ambient temperature (25°C) from the maximum temperature (175°C) and then divide the result by the junction-to-ambient coefficient maximum listed on the datasheet. (Ambient temperature is the temperature of the air around something, a junction is where two parts come together, and coefficient simply means a factor.) As long as the power calculated for the transistor is less than that, no heat sink is needed.

Unfortunately, the datasheet found for this particular transistor doesn't list the junction-to-ambient coefficient maximum; however, it does show the maximum power dissipation of 96 W, so compared to the 0.00423 W that the circuit will produce, it should be just fine.

Building the Circuit

Now let's build the circuit. The sample circuit was created using a small slider switch. A push button could also be used as long as it is not a momentary type. The following instructions are for the demo breadboard. You may have to adjust your pin locations accordingly. Figure 9.11 shows the circuit schematic and breadboard connections to assist in constructing the circuit. Figure 9.12 shows the completed circuit.

1. Ensure that the switch is in the off position and place the switch such that the pins being used are in 18d and 19d. For most switches with three pins, you will use one outside and the center pin.

2. Place the transistor so that the gate is in 27d, the drain is in 26d, and the source is in 25d.

3. Jumper from the gate to the outside pin of the switch (27c to 18c).

4. Place the ends of the 10k-ohm resistor in 27a and the ground rail.

5. Jumper from the source pin, 25a, to the ground rail.

6. Connect the negative lead of the load to the drain at pin 26a.

7. Connect the positive lead of the load to the positive rail.

8. Connect the 12V positive lead to the positive power rail.

9. Connect the 12V negative lead to the ground rail.

10. Connect the 6V positive lead to the center pin of the switch.

11. Connect the 6V negative lead to the negative power rail.

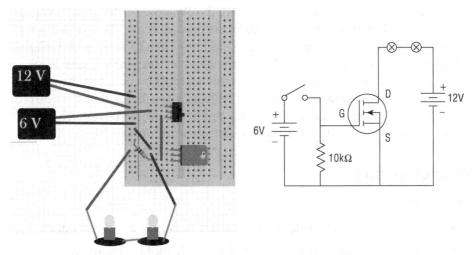

Figure 9.11: A circuit schematic

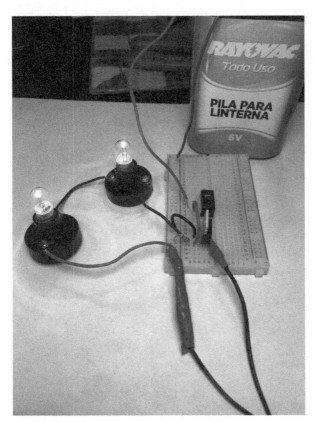

Figure 9.12: A transistor switch circuit

When the switch is slid to the on position, the 6V source will supply a voltage to the gate, which will enable a larger current to flow from the 12V source through the load and back to the circuit ground. Slide the switch back off and the lights will go off again, dissipating any leftover voltage on the gate through R1.

It might seem like this could be done with a simple switch, but the point to remember is that the smaller voltage can be from a source such as the one created with an Arduino or computer output, and it can be used to switch a much larger voltage, in the case of this particular transistor, up to a 60V load.

Troubleshooting

Here are some troubleshooting tips:

- If the circuit doesn't work, use Figure 9.11 to help check the connections.
- Verify that the transistor is oriented properly (not backward).
- If the circuit doesn't turn back off when the switch is turned off, check the resistor connections.
- If the circuit doesn't turn on, check the power connections and the battery voltage rise.

Capacitors

"The willingness to show up changes us; it makes us a little braver each time."

—Brené Brown

Why the quote about bravery? One of the circuits here might be a bit difficult to understand, like why some people have a hard time understanding that gender doesn't determine ability to do anything. I don't know if you're a guy or a girl reading this, but trust me, you've got this. I'm here to help you along the way.

If you've ever had the opportunity to use a shake flashlight, you might wonder how it holds the electricity generated. One way is with a small supercapacitor. Often capacitors (*caps*) are described as being similar to batteries because they can hold a charge, but they have many more uses than that. In this chapter, we delve into some of those uses and see how much fun working with capacitors can be.

The unit for measuring a capacitor's ability to hold electrons is the *farad*. One farad is defined as the ability to hold a charge of one coulomb at a potential difference of one volt. The farad is a rather large unit, so capacitor values are generally expressed in pico farads (0.000,000,000,001 F = 1 picofarad) or micro farads (0.000,001 F = 1 microfarad).

A Quick Look at Capacitors

Capacitor construction is fairly simple. In general terms, capacitors consist of two metal plates separated by a *dielectric* material. Dielectrics are insulators, meaning they don't easily give up electrons. One of the plates of the capacitor is charged with electrons. The dielectric material keeps the two plates separate. When a connection is made between the two metal plates, the charges rapidly equalize, causing a sudden flow of electrons. Figure 10.1 shows the components of a simple electrolytic capacitor.

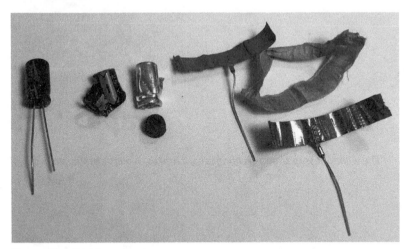

Figure 10.1: Inside a capacitor

Before building a circuit with capacitors, it's necessary to know which one to choose. There are many different types of capacitors; Figure 10.2 shows a few. Ceramic disc capacitors look like the name implies, like a small round disk. Disc capacitors usually have a number stamped on them indicating their value. They can also have other information stamped on them, such as letters to indicate tolerance, maximum voltage, or heat. Determining the value of a capacitor marked in this way is similar to determining the value of a resistor, but capacitors are in picofarads. If, for example, a disc capacitor is labeled 474, the first number is a value, the second number is a value, and the third number is a multiplier, meaning to add that many zeros to the value, so the capacitor marked 474 would be 470,000 picofarads, or 0.47μF (0.47 microfarad).

Disc capacitors typically don't have polarity, and neither do plastic film capacitors. Electrolytic and tantalum capacitors typically do have polarity. If a capacitor requires specific polarity, the negative side will be indicated by something different like a stripe on one side or the typical shorter leg. If a capacitor has polarity, it's imperative that it is connected properly. Failure to connect it with the right polarity may result in the dielectric material failing or, worse, the capacitor swelling or exploding. Figure 10.3 shows a failed capacitor on a motherboard. Even if nothing that dramatic happens, it will likely be damaged and have to be replaced.

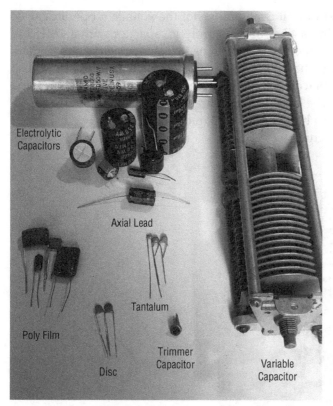

Electrolytic
Capacitors

Axial Lead

Tantalum

Poly Film

Disc

Trimmer
Capacitor

Variable
Capacitor

Figure 10.2: Common capacitor types

Plastic film caps (or polyfilm caps) are named after the dielectric material within them, such as polyester, polypropylene, polytetrafluoroethylene (PTFE [whose most well-known brand name is Teflon]) film, or polystyrene film. They are typically flattened, are rectangular, and look shiny, rather like a piece of candy-coated gum. Some simply look like a small plastic box. These may also have numbers and letters on them to indicate values and tolerances.

The capacitors that look like small cans are electrolytic capacitors. If the leads extend from each end instead of the bottom, they are called *axial lead capacitors*. Electrolytic capacitors do have polarity, and it must be observed. The hash marks (X) on the top of the electrolytic capacitors are there to control the direction of explosion, should it happen. When troubleshooting a circuit, a swollen capacitor or one that is oozing its contents is a simple problem to visually observe, and the capacitor must be replaced.

When replacing a capacitor, the best course of action is to replace it with an identical capacitor. If a particular thermal limit is needed, be sure to observe that. Be sure to meet or exceed the capacitor's specified *voltage*. The voltage marking on a capacitor is the maximum it can handle, not the voltage it operates at, so in most instances, replacing a 47uF, 35V electrolytic capacitor with a 47uF, 50V capacitor should work just fine. It simply means that the capacitor replacement can handle a higher voltage than the circuit will give to it. It's important to keep the farad rating of the replacement capacitor the same as the capacitor being replaced. Remember that the farad is a measure of the power stored in a capacitor, so replacing one with a higher or lower farad rating could have disastrous consequences for your circuit.

Figure 10.3: Failed electrolytic capacitor

Variable capacitors are also known as *tuning capacitors* or *trimmer capacitors*, depending on their size and construction, and they have the ability to change the amount of energy they can store by changing the amount of overlap of moving plates. When you tune an older radio by turning a knob, you're likely adjusting the capacitance of a variable capacitor. (Many radios now are on an integrated circuit chip, which can be controlled by software.) The symbol for a variable capacitor has an arrow to show that it can be changed, or an arrow with a bar across it like the letter "T" for a trimmer cap. Trimmer capacitors are often tiny capacitors that are typically found soldered onto a circuit board and can be adjusted using a plastic tool that looks like a tiny slotted screwdriver. They're often used for initial adjustment of a device or for fine-tuning. Figure 10.4 shows symbols for various types of capacitors. A curved line or blackened box is often used to indicate the negative side of a polarized capacitor. Variable capacitors can also have polarity or not.

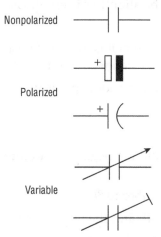

Nonpolarized

Polarized

Variable

Figure 10.4: Capacitor symbols

Try This: Creating a Time Delay Circuit

Capacitors have the ability to hold a charge, which they will rapidly discharge when a circuit is completed from one terminal to the other. In an oversimplified example, that's how the flash on a camera works; the high-voltage electricity is discharged in an instant through the camera's flash circuit when the shutter button is pressed. The rate of discharge of the capacitor can be controlled by putting a resistor in series with the capacitor, as the following circuit demonstrates. This type of circuit is called an *RC circuit*. For this project, you'll need the following materials:

- Small breadboard
- Slider switch
- LED
- 1000uF electrolytic capacitor
- 330-ohm resistor
- 9V battery with leads
- Assorted jumpers

Optional:

- 2200uF capacitor
- 10uF capacitor

The following instructions reference the positions on the breadboard used in the demo circuit. The breadboard you use may be slightly different. Use Figure 10.5 to help in connecting your circuit.

1. First, test the switch with an ohmmeter to ensure that it is in the off position. (The resistance between the middle and end lead will be infinite.)

2. Place the switch so that the end pin is in 4h and the middle pin is in 5h.

3. Jumper from the positive rail to 5j.

4. Jumper from 4f, across the dip to 4e.

5. Connect the positive lead of the LED into 4d and the negative lead into 6d.

6. Connect one end of the resistor into 6e and the other to 8e.

7. Connect the positive lead of the capacitor into 4b and the negative lead into the negative rail.

8. Jumper between 8a (the second lead of the resistor) to the negative rail.

9. Connect the leads of the battery into the positive and negative rails, respectively.

10. Flip the switch on. The LED lights immediately.

11. Measure the voltage across the capacitor leads. It should be close to the voltage rise across the battery.

12. Now, turn the switch off.

The LED stays lit, fading slowly even though there is no power source connected to the circuit (because the switch is open). The power is being supplied by the capacitor (see Figure 10.6). The instant that the circuit is turned on, the voltage is at zero. The current rises rapidly. The LED and capacitor are in parallel to each other, so each is receiving the full voltage of the source. But when the potential difference between the two plates of the capacitor reaches the source voltage, current no longer flows into the capacitor branch, and the LED branch acts like a series circuit. When the switch is opened, the capacitor and LED form essentially a series circuit, and the capacitor lights the LED until its energy is depleted.

Try replacing the capacitor with a 2200uf capacitor and then a 10uf capacitor. In each case, if the voltage across the capacitor is measured with the switch on, it will be close to the voltage rise, but observe the difference in the time it takes for the LED to go out when the switch is turned off.

Figure 10.5: Time delay circuit

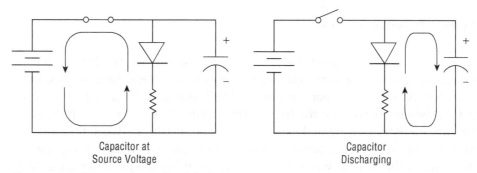

Capacitor at
Source Voltage

Capacitor
Discharging

Figure 10.6: Capacitor charged and discharging

> **NOTE** Capacitors aren't perfect devices, and some leakage will occur through the dielectric. Eventually, a capacitor will discharge on its own, but even when intentionally discharged, a high-voltage DC capacitor can have electrons that are trapped in the dielectric. When those electrons make their way out of the dielectric, they essentially recharge the capacitor to some extent, so always approach large capacitors with caution. To safely discharge a capacitor, carefully connect its leads across an appropriate resistance.

For a moment, ignore the role of the diode in the circuit and focus on the resistor. Whenever a resistor is in series with a capacitor, it slows down the current whether it's being discharged from or "filling" the capacitor. The greater the resistor and capacitor values, the longer this time becomes. The time is called the *transient response* (Figure 10.7). A transient response requires five blocks of time called *time constants* (5τ) to complete. Each time constant fills or depletes the energy in the capacitor by 63.2%. τ is the time in seconds, and τ = RC where R is resistance in ohms and C is capacitance in farads. τ is a single time constant, and it takes five of them to equal the transient response.

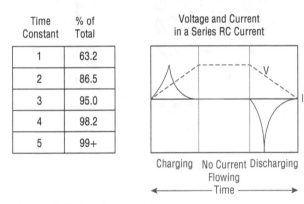

Time Constant	% of Total
1	63.2
2	86.5
3	95.0
4	98.2
5	99+

Voltage and Current in a Series RC Current

Charging   No Current Flowing   Discharging

Time

Figure 10.7: Transient response

Again, ignoring the diode, with a 330-ohm resistor in the circuit and a 1000uF capacitor, the time constant would be τ = 330ohms*0.001F, for a time constant of 0.33 seconds × 5 = a transient response of 1.65 seconds. The transient response is used where controlling the time to turn a circuit on or off is important. In the time-delay circuit created earlier, there is no resistor between the capacitor and the source, so the transient response is not observed when the capacitor is charging. If the resistor were moved to the main branch just after the switch, it could be.

Suppose a circuit was needed that would turn on an alarm 30 seconds after a door was opened or when someone walked past a motion sensor. Thirty seconds is the desired transient response. The transient response divided by 5 means that the time constant would need to be 6 seconds. If a 1000uF capacitor were used on the circuit, the circuit resistance would need to be 6000 ohms (6 seconds divided by 0.001 F). If combined with an amplifying transistor (see Chapter 9, "Transistors") or a relay (see Chapter 11, "The Magic of Magnetism"), almost anything could be turned on or off by a time delay circuit.

A graph of the transient response would look similar to Figure 10.7. One line represents voltage, and the other line represents current. With a capacitor in a circuit, the current flow will rapidly "fill" the capacitor with electrons until the voltage measured across the capacitor is very close to the source voltage. Once the capacitor reaches this point, no more current will flow through the circuit. When the source is removed, the capacitor remains charged until it has a place to discharge, meaning something completes the path between the capacitor's two connectors (plates).

Capacitor Uses

Capacitors are quite useful components. As in the previous example, they're used in timing circuits, but they serve many more functions.

Capacitors can be used to block a DC signal but pass an AC signal. This happens because the capacitor charges and discharges quickly as the AC signal rises and collapses. DC voltages will go to their peak and stay there, so the capacitor appears to block a DC voltage and pass an AC voltage. Following are some of the ways capacitors are used:

- Noise filter circuits
- Smoothing rectified (AC to DC) outputs, such as in a DC power supply
- Signal coupling and decoupling (separating DC from AC)
- Remote sensing, such as in building integrity and telecommunications
- Tuning to accept a specific radio signal
- High-pass and low-pass filters (block and allow certain frequencies or bands of frequencies, for example, woofer and tweeter speakers)
- To start and stop motors quickly
- Line conditioning (absorb power fluctuations)
- Flashing lights (timer circuits again)
- Computer memory
- Lasers
- Oscillator circuits

There are so many uses of capacitors that it would be difficult to go through a day in our modern world without using a capacitor, even if unaware.

Try This: Creating an Astable Multivibrator

What geek girl doesn't love flashing lights? Astable means not stable, so as implied this circuit does not maintain a stable state. It uses capacitors and transistors to switch circuits on and off alternately, causing the connected lights to flash. It could just as easily be done with buzzers replacing the LEDs. For this project, you will need the following materials:

- Small breadboard
- (2) LEDs, any color
- (2) BC547B or similar NPN transistors
- (2) 100uF capacitors
- (2) 330-ohm resistors
- (2) 33k-ohm resistors
- 9V battery with leads
- Assorted jumpers

Optional:

- (2) 10uF capacitors
- (2) 220uF capacitors
- Assorted resistors

The following instructions reference the positions on the breadboard used in the demo circuit. The breadboard you use may be slightly different. Use Figure 10.8 to help in connecting your circuit.

1. Insert the negative lead of one capacitor in 11c and the positive lead in 8c.
2. Insert the negative lead of the other capacitor in 11i and the positive lead in 8i.
3. Insert the negative lead of one LED in 8d and the positive lead in 6d.
4. Insert the negative lead of the other LED in 8j and the positive lead in 6j.
5. Insert a 330-ohm resistor between 6a and the positive rail connecting the positive side of the LED to the positive rail.
6. Insert the other 330 ohm resistor between 6k and the positive rail on that side of the board connecting the other LED to the positive rail.

7. Insert a 33k-ohm resistor between 11a and the positive rail, connecting the negative side of a capacitor to the positive rail.

8. Insert the other 33k-ohm resistor between 11k and the positive rail on that side of the board.

9. Insert a transistor so that the emitter is in 20b, the base is in 19b, and the collector is in 18b.

10. Insert the other transistor so that the emitter is in 20k, the base is in 19k, and the collector is in 18k.

11. Place a jumper between 20a and the negative power rail, connecting the transistor's emitter to the negative side of the circuit.

12. Place a jumper between 20l and the negative power rail, connecting that transistor's emitter to the negative rail on that side of the breadboard.

13. Jumper from the negative side of each LED (8e and 8h) to the collector of the transistor on that side of the board (18e and 18h).

14. Jumper from the negative side of the first capacitor, 11g, to the base of the transistor on the opposite side of the board, 19f.

15. Jumper from the negative side of the other capacitor, 11f, to the base of the transistor on the other side of the board, 19g. *Note: typically it's best to avoid wires crossing over each other on the board, but it is done here for simplicity's sake.*

16. Finally, place a jumper between the two positive rails and another jumper between the two negative rails of the breadboard so that the positive rails are connected and the negative rails are connected.

17. Look closely at your breadboard. If there is a gap on the long red line or the long blue line, then the rail on top may be separated from the rail on the bottom (see Figure 10.8). If so, place a small jumper to connect the sides of the positive or negative rails as needed. The demo board needed such a jumper on the positive rails of the board.

18. Verify component connections and then connect the power source to the positive and negative rails, respectively.

The LEDs should start flashing alternately.

Why does this circuit work this way? First, the 330-ohm resistors are there only to limit current to the LEDs. Notice their position on the schematic in series with the LEDs. The 33k-ohm resistors work with the capacitors to control the speed at which the circuit oscillates. Oscillating simply means that something is going first in one direction and then the other in a rhythmic sort of motion. Think of a musician's metronome or the pendulum of a grandfather clock.

Refer to the schematic in Figure 10.8. The transistor Q_1 will turn on when the voltage at the base of Q_1 is 0.7 volts or more, and because the base is connected to C_2, C_2 effectively controls Q_1; likewise, C_1 controls Q_2.

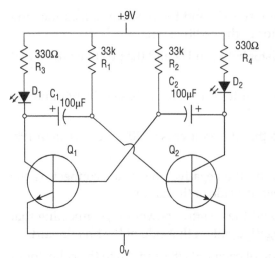
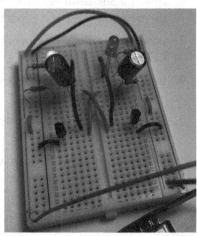

Figure 10.8: Astable multivibrator circuit

Even if they have the same part number, no two components, or circuits for that matter, will be exactly alike, so when the circuit is turned on, one transistor will start conducting before the other. Assume that Q_1 is conducting.

When Q_1 is on, a small current is flowing through the base of Q_1 and its emitter to the negative side of the circuit. A positive voltage of 0.07V or greater exists at the base of Q_1, which has "turned on" the transistor and allowed a larger current to flow through R_3 and D_1 and Q_1's collector and emitter to the negative side of the circuit. The voltage at the collector of Q_2 is near the source voltage because Q_2 is acting as an open switch. No current can flow through R_4 and D_2. C_2 will be quickly charged to approximately 8.3V (9 – 0.7V). C_1 will be charging through R_1 with its charging time determined by the values of C_1 and R_1 (($\tau = R_1C_1$)*5). C_1 never gets to reach a full charge because when the voltage of C_1 reaches 0.7V, Q_2 turns on.

When Q_2 turns on, the voltage at Q_2's collector drops quickly as Q_2 conducts through its collector and emitter. D_2 turns on as current flows through it to Q_2. The sudden drop in voltage at Q_2's collector pulls down the voltage at C_2, and the voltage at the base of Q_1 drops to about -8.1 volts, turning off Q_1 and D_1. The voltage at Q_1's collector is now approximately 9V because Q_1 is acting as an open switch. C_2 begins to charge through R_2. C_1 is charging through R_3, but not enough to turn on D_1. C_1 will charge to about 8.3V.

When C_2 gets up to 0.7V, it will again turn on Q_1, and Q_2 turns off. The process repeats until the source voltage is disconnected.

I know. You may have to read that more than once for it to sink in.

Because the speed at which the capacitors charge is determined by the values of each resistor/capacitor pair (R_1 and C_1, R_2 and C_2), the rate of flashing can be changed by changing the resistors and capacitors. That's where the optional components in the materials list come in.

Before continuing, you might want to add a slider switch to your circuit.

1. On an unused area of the circuit board, insert the slider switch. In the demo board, its pins are in positions 26d, 27d, and 28d.

2. Insert a jumper between the center pin and the positive rail (27a to positive rail).

3. Insert the positive lead of the battery in 28a (same row as the center switch pin).

4. Insert the negative battery lead into the negative rail.

When you turn the switch on and off, it should now turn the circuit on and off. This will make changing the capacitors and resistors easier because the circuit can be turned off with the switch instead of disconnecting a wire. Figure 10.9 shows a switch added to this circuit.

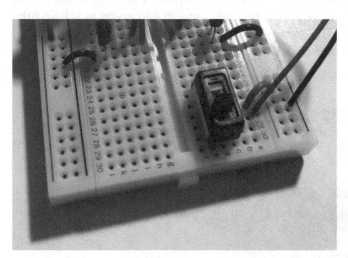

Figure 10.9: Inserting a switch

Try changing out the capacitors for higher and lower values, perhaps not even the same value on each side. Change out the resistors and see how that affects the circuit. It's one thing to read that a change in capacitance and resistance will affect the speed, and quite another to experience it, so please, have fun and experience controlling the flow of electrons, but leave the circuit together. It will be used in the next section.

Try This: Using Capacitors in Series and Parallel

Capacitors in series and parallel with each other may behave differently than you'd expect. Assume you need 220uF of capacitance and all you have are two

100uF capacitors and two 10uF capacitors. How would they be connected to equal 220uF? You're about to learn.

For this project, you'll need the following materials:

- The astable multivibrator circuit from the previous project with two 100uF capacitors inserted

- An additional 100uF capacitor

Capacitors in Parallel

The following instructions reference the positions on the breadboard used in the demo circuit. The breadboard you use may be slightly different. Use Figure 10.10 to help in connecting your circuit.

1. Place a 100uF capacitor in parallel with a capacitor already in the circuit by placing the negative end of the new capacitor in position 11k and the positive end in 8k. The negative leads of both capacitors will be in row 11, and the positive leads of both capacitors will be in row 8.

2. Turn the circuit on.

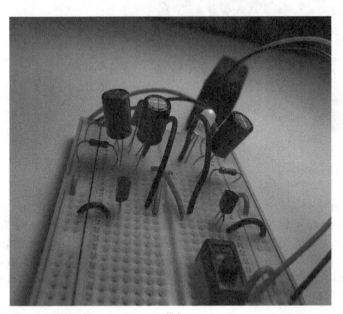

Figure 10.10: Capacitors in parallel

What do you observe about the flashing rate of the LED with the capacitors in parallel? It should stay on longer, which means it is discharging more slowly, so therefore, the capacitance is increased. That's right! Capacitors in parallel increase capacitance.

When capacitors are in parallel, the math is exceedingly simple. The equivalent capacitance is equal to the sum of the capacitors that are in parallel. The following formula applies, where C_N is the highest numbered capacitor of any number of given capacitors:

$$C_T = C_1 + C_2 + ... C_N$$

Capacitors in Series

Continuing with the earlier activity, follow these steps:

1. Turn the circuit off and remove the capacitor you just inserted in step 1 of the previous section.

2. Remove the LED on that same side.

3. Remove the 330-ohm resistor for that LED.

4. Place the second capacitor so that its negative lead is in the same row as the positive lead of the first capacitor. On the demo board, position 8j. Insert the positive lead in position 4j. Insert the negative lead of the LED in position 4i and the positive lead in position 2i.

5. Insert one end of the 330-ohm resistor in 2k and the other in the positive rail.

6. Turn the power on.

The LED will likely flash just briefly; then the opposite LED appears to stay on. But if you watch closely, you may observe that the LED with the capacitors in series flickers on faintly from time to time and then goes out. This happens because the capacitance on that side is so low that it isn't enough to keep the LED glowing.

Capacitors in series lower the equivalent capacitance. The total capacitance will be lower than any one of the capacitors. Like resistors in parallel, there are three ways to calculate the value of capacitors in series.

1. When there are more than two capacitors of unequal values, and n is the highest numbered capacitor of any number of capacitors:

$$\frac{1}{C_T} = \frac{1}{C_1} + \frac{1}{C_2} + ... \frac{1}{C_N}$$

2. Where all capacitors have the same value, the total capacitance is the capacitance of 1 divided by the number of capacitors. For example, three 1200uF capacitors in series would have an equivalent capacitance of 400uF (1200/3).

3. If there are two unequal capacitors, the product over sum formula can be used.

$$C_T = \frac{C_1 \times C_2}{C_1 + C_2}$$

So, the answer to the question at the beginning of this section? The capacitors need to be in parallel.

CHAPTER 11

The Magic of Magnetism

"Well-designed goals are like a magnet—they pull you in their direction and . . . the harder you work on them, the stronger they pull."

—Jim Rohn

When you were a kid, did you ever have a one of those funny, magnetic face characters where you used a "magic" pen to draw eyebrows, a beard, and a mustache on the face? I was surprised to see that they're still on the market, but they're a lot of fun. This is magnetism at play, but magnetism performs serious work, too. Without the magic of magnetism, I likely wouldn't be writing this book on my computer, because large-scale electricity as we know it likely would not be. Magnetism is used in power generation, yes, but in lots of other ways, too.

Magnetism is used to sort single-stream recyclables, remove contaminants from food and medicine, and treat patients, both in ancient medicinal practices and in making our modern drugs. Magnets are instrumental in running a computer's magnetic hard drives. Large electromagnets are used on construction sites. Magnetic resonance imaging (MRI) uses strong magnets in the imaging process. Magnets are often used to seal doors and safes. Magnetism is what moves a solenoid, like the electronic lock on your car door. They're also used in speakers and microphones, and too many other places to list here. Read on and explore just a few of the ways that we put magnetism to work in modern electronics.

The Electricity/Magnetism Relationship

To review the relationship between electricity and magnetism, we first must discuss magnetism.

Magnetism

The first question to answer is, what is magnetism? *Magnetism* is an invisible force that exists around certain objects, called *magnets*. Just like electricity has an electromotive force (EMF), a magnetic field's push is called *magnetomotive force (MMF)*. The area encompassed by this force is called a *magnetic field*, and it will attract only certain metals. Iron, nickel, and cobalt are naturally occurring *ferromagnetic* materials, meaning they are easily magnetized. Most material that is *ferrous* (i.e., containing iron) is attracted to a magnet, although not all alloys are.

Magnetism is caused by the molecular alignment of atoms in a material such that at any given time the positive and negative forces of a group of atoms are aligned with each other. These groups of aligned atoms are called *magnetic domains*. See Figure 11.1. In a magnet, the magnetic domains are aligned in the same direction. A common screwdriver can be turned into a magnetic screwdriver by stroking it with a magnet in a single direction to align the magnetic domains. Ferromagnetic materials can be turned into magnets by subjecting them to a strong magnetic field for a time. *Lodestones*, which are made of the mineral magnetite, are naturally occurring magnets and have been used for thousands of years. *Neodymium* is a rare Earth metal that is used to make some of the strongest magnets on Earth.

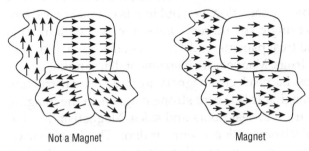

Not a Magnet   Magnet

Figure 11.1: Magnetic domains

Magnets have north (north-seeking) and south (south-seeking) poles just like Earth, which is itself a large magnet due to all the molten iron in its core. Magnetic poles behave similarly to opposite charges, so opposite poles attract (north to south), while poles that are the same repel each other. Compasses work because they have a needle made of ferromagnetic material, and the compass needle's south pole will always point toward Earth's magnetic north pole, which, incidentally, is located at what we call the South Pole. A compass's north pole will point in the direction we call the North Pole because magnetically it is south, even though we call its location magnetic north.

Magnetic north (i.e., the North Pole) and geographic north are not the same place on Earth. In fact, over time magnetic north may shift and can vary from "spot on" the geographic north to hundreds of miles away. We call this difference the *angle of declination* or the *angle of variation*, and it must be taken into consideration in applications (i.e., circuits) where magnetic direction is used. The angle of declination also depends on where on the globe it is being measured from.

Look again at Figure 11.1. In a magnet, the magnetic domains are aligned in the same direction, so if a magnet is broken, instead of having one piece that is south and the other that is north, there will simply be two magnets, each with its own north and south poles. The molecular alignment remains the same.

NOTE   Weaker magnets can be demagnetized by dropping them or subjecting them to heat. Store and handle magnets carefully to preserve their magnetic quality.

The lines of force in a magnet are called magnetic lines of *flux*. They form a stronger force at the poles of a magnet where they are concentrated, and they always extend from the magnet's north pole to the south pole. Figure 11.2 illustrates this, but remember that it is a two-dimensional representation of a three-dimensional force that would be 360 degrees around the magnet. The more lines of flux, the stronger the magnet is. Magnetic strength is measured in a unit called a *weber* (Wb). Think of the weber like electrical current. 1 Wb/square meter is called a Tesla (T), so named after Nikola Tesla, the genius inventor. So, the Wb is the flux quantity, and the T is the unit for how tightly packed together it is. The symbol B stands for flux density, and the formula for flux density is $B = \dfrac{\Phi}{A}$, where Φ (the Greek letter phi, pronounced f-eye) is the number of lines of flux (in webers) and A is the area (in square meters). If the flux is 100Wb and the area is 100 square meters, then B = 1T.

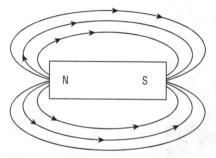

Figure 11.2: Magnetic lines of flux

Magnetic lines of flux can travel through anything, but they travel through some materials more easily than others. *Permeability* (μ—yes, the same symbol as micro) is the ability of magnetic flux lines to traverse a material. Conversely, *reluctance* (S) opposes a magnetic field like resistance opposes EMF. Air has a relative permeability of just over 1, while iron can have a permeability of thousands, depending on how pure it is. Knowing the permeability of a material can be important to the electronics inventor's work, because using permeable materials enables us to direct magnetic forces where we want them to go. Like electricity, magnetic fields take the path of least resistance, so a more permeable material, which is less resistant to magnetic lines of flux, can act as a shield protecting other devices from the magnetic field.

Magnetism's Relationship with Electricity

The relationship between electricity and magnetism is a yin-yang sort of relationship. If you take a small spool of hookup wire and bare both ends, then connect those ends to a galvanometer (a tool for measuring very small currents), and finally run a magnet back and forth in the middle of the spool, you'll see the galvanometer's needle move first in one direction and then in the other. This proves the point that when there is a magnetic field, a conductor, and relative movement between the two, the magnetic field will induce a current to flow in the conductor.

This same principle is how much of our electrical grid power is generated. Whether the mechanical movement is water-powered turbines from the Hoover Dam or wind-powered turbines like the wind farms on the Tug Hill Plateau of New York (Figure 11.3), the principle is the same. A magnetic field and mechanical movement cause a current to flow in conductors that are intersected by the magnetic field, and that current is our power.

Figure 11.3: Wind farm

But wait! There's more! Anytime a current is flowing through a conductor, two things happen. Some energy is lost as heat due to the resistance of the conductor, and a magnetic field is created around the conductor. What? Yes, it's true. Current flowing through a conductor causes a magnetic field around the conductor. This is how the huge electromagnets on construction sites work. Speaking of electromagnets, let's make one.

Try This: Building an Electromagnet

I will admit that this lab is a bit more fun than substance, but it illustrates the ability of a conductor to become a strong magnet. This project uses magnet wire, which is finer than the hookup wire used in most projects and is coated with shellac instead of a thick plastic coating. For this project, you will need the following materials:

- Steel framing nail or decking screw
- 9V battery
- (3) conductors with alligator clips on each end
- 10 feet of 30-gauge magnet wire
- 10-ohm, 10W resistor
- Paper straw or small piece of paper and tape
- Small piece of sandpaper
- Handful of small, uncoated metal paper clips, or similar objects

WARNING Leave the circuit connected for only a few seconds. Longer may cause excessive heat and finger burns. If it smells or feels hot or the battery gets warm or swells, disconnect it immediately.

Follow these instructions to build your electromagnet:

1. Cut a piece of paper or paper straw slightly shorter than the nail. If using paper, wrap it once around the nail loosely enough so you can slide it off and then tape it.
2. Place the paper or straw around the nail.
3. Measure approximately 10 feet of the magnet wire.
4. Leaving 2 inches of the magnet wire sticking out past one end of the nail, begin wrapping the magnet wire around the nail with the turns close together, distributing the wire evenly along the length of the nail. End at the opposite end from where the wrapping started, and tape the second wire end so approximately 2 inches of it are sticking out.

5. Use the sandpaper to remove the shellac from about 1/2 inch on both ends of the magnet wire.

6. Place the small metal objects on a surface close by.

7. Clip one end of an alligator clip to the positive side of the battery, and clip the other end to one lead of the resistor.

8. Clip another alligator clip to the other lead of the resistor and the other end to one end of magnet wire that is wrapped around the nail.

9. Clip a third alligator clip between the other end of the magnet wire and the negative side of the battery. Your electrical circuit should be a loop from the battery through the resistor and the magnet wire back to the battery.

10. Move the nail near the metal objects, and the electromagnet should pick them up!

Current flowing through the conductor created a magnetic field. The iron core adds to the strength of a magnetic field. Other ways to increase the strength are to increase the number of turns of wire, wrap the strands of magnet wire more closely together, or increase the current. See Figure 11.4.

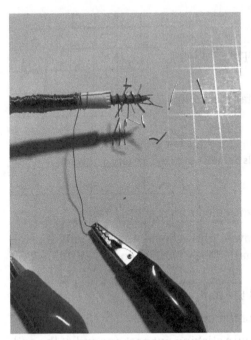

Figure 11.4: Electromagnet

Magnetism in Circuits

Not only are magnetic fields used for electromagnets as demonstrated in the previous experiment, but they are used in circuits for numerous other purposes.

As with all things in electronics, there are some rules for magnetism in circuits. First, the left-hand rule for conductors: if you grasp a conductor with your left hand and your thumb points in the direction of current flow, the fingers are curled around the conductor in the same circular direction as the magnetic field.

When a current-carrying conductor is coiled as it was in our electromagnet, the magnetic fields of the coils are combined, and the result is a stronger magnetic field. With your fingers wrapped around the coil in the direction in which the current is flowing, your thumb will point to the north pole. This theory is easy to test with your electromagnet. Place a compass near one end of the energized electromagnet. If you are at the north pole of the electromagnet, the compass's south arrow will be pointing toward it. (Opposites attract.) Change the direction of the current (i.e., reverse the positive and negative connections on the battery), and the poles will change.

In Figure 11.5, the first image's center arrows show the direction of current through a conductor, and the arrows around the outside show the directional rotation of the magnetic field generated around those conductors. The second image shows a coil with the current direction indicated with small arrows and the north and south poles identified. It would also have magnetic lines of flux from the north pole to the south pole all around the coil. The third image shows the combined magnetic field of a conductor's coils. The image shows two coils that have current flowing through them in the same direction. Each has a magnetic field around it, which is the combination of its coils' magnetic fields. With the current moving in the same direction, the south pole of one coil is attracted to

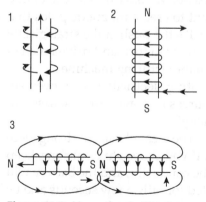

Figure 11.5: Magnetism in Circuits

the north pole of the other coil. As the two coils come together, their magnetic fields will join into one larger, more powerful magnetic field. Remember, this is a two-dimensional drawing of a three-dimensional object, and the lines of flux would be around the entire coil. If the two coils are near each other but not physically connected, a changing current in one will cause a similar reaction in the other. This is called *mutual induction*. Mutual induction will be discussed further in Chapter 16, "Transformers and Power Distribution".

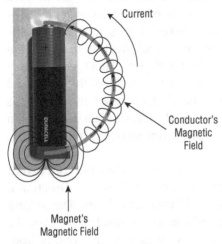

Figure 11.6: Homopolar motor

Motors and Generators

Motors and generators take advantage of the fact that magnetic fields have north and south poles, which repel or attract each other (same poles repel, opposite poles attract) just as electrical charges do. Motors turn electricity into mechanical motion by using electrical current to create magnetic poles that oppose the magnetic poles on a rotating piece. It's the push of the strategically placed magnetic poles that causes the motor's rotor to turn and drive much of our world, from toys to sewing machines to manufacturing machines.

Generators are the opposite of motors, generating electricity from movement between conductors and magnetic fields. Sources of power are discussed in more detail in Chapter 15, "Sources of Electricity".

A homopolar motor is a simple motor that can be created using a strong (neodymium) magnet, a battery, and a conductor. If you choose to create one, keep in mind that there is very little resistance in the conductor, and the battery will overheat. While it's fun to do, leave it connected for only a few seconds and be mindful of the possibility of burning your fingers on the conductors or damage to the battery due to the battery overheating. See Figure 11.6. The hardest part of making a homopolar motor is getting the conductor perfectly balanced.

The conductor has a pivot point on the top of the battery, and on the bottom it touches the neodymium magnet. The magnet provides one magnetic field. The electrons rushing from the negative end to the positive through the conductor provide the opposing (same) magnetic field, causing the conductor to rotate around the battery.

Inductors

An *inductor*, which is essentially a coil of wire, is used in a circuit to store a small amount of energy that is returned to the circuit when the power is disconnected. They can also be used to transfer energy from one circuit to another through *mutual induction*. When two inductors are in close enough proximity with each other that the magnetic field of one crosses the conductors of the other, the rising and falling of the magnetic field in one will induce a current to flow in the other. This is mutual induction. If the current is DC, then the magnetic field in a coil will rise and stay at a fixed point. When there is no movement (i.e., no rising or collapsing magnetic field), there is no more current induced in the second coil. So, quickly after the DC current reaches its peak and stays constant, it no longer induces a current in the second coil because there is no movement between the magnetic field and the conductor. If, however, the current in the first coil is AC, the magnetic field rises and collapses repeatedly, providing the movement between coil and magnetic field that induces an AC current in the second coil via mutual induction.

Inductors are like capacitors in that they hold energy, and like capacitors, inductors have a transient response. It takes five time constants for an inductor's transient response, just like a capacitor. The difference is that with inductors, it's the current that rises slowly and stays steady in the circuit until there is a change in the circuit's voltage.

Air core inductors are typically used for high-frequency implementations, while iron core inductors are used for lower frequencies. If iron core inductors were used with high frequencies, the constant switching of direction would cause excessive molecular friction and heat, which also means lost energy. Air core inductors may have a plastic or other core that is not ferromagnetic material and is used simply to support the coil of wire. Inductors can be fixed or variable and can even be "tapped" at different points along their length, in which case they are essentially autotransformers. Figure 11.7 shows common schematic symbols for inductors.

NOTE Frequency refers to the number of times per second an AC signal will go from zero to positive peak and then negative peak and back to zero, which is also known as one complete cycle. Frequency is measured in cycles per second, also called hertz (Hz). If a signal has a frequency of 60Hz, then it completes 60 cycles every second.

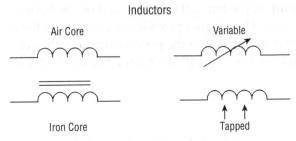

Figure 11.7: Inductor schematic symbols

Inductance, which is essentially the inductor's ability to create a magnetic field and hold energy in it, is measured in a unit called *henrys*, often expressed as millihenries. Many texts define inductance as a tendency to resist a change in current, which the inductor does by slowing the rise of circuit current when a circuit is turned on and supplying current back into a circuit as the magnetic field collapses (i.e., the transient response), such as when power is turned off or when AC flows in the opposite direction back toward zero.

Inductors can be found in devices like radios or low-pass LC circuits, or smoothing out the voltage in a power supply. They're also found on computer circuit boards. The circuit board in Figure 11.8 has at least two inductors on it. There may be more. In Chapter 16, we'll look at how inductors are used to create high voltages. Not all inductors look like the ones on the circuit board. Some inductor packages look like small plastic canisters or boxes, and others look a bit like elongated resistors, just to name a few.

Figure 11.8: Inductors on a motherboard

Doorbells

An old-fashioned doorbell is a clever use of magnetism. It consists of a simple circuit with a coil to create a magnetic field and a metal bar that will move. See Figure 11.9.

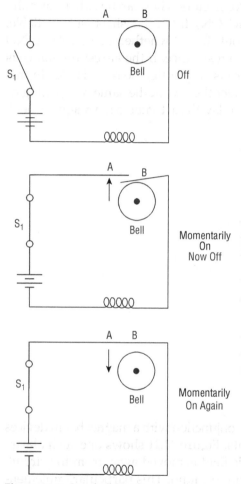

Figure 11.9: Doorbell circuit

When the button (switch) is off, there is no magnetic field, and the metal bars at point A and B are connected in the circuit like a closed switch.

When the button is pressed, S_1 closes, and the circuit is complete, so electricity flows, and the magnetic field is energized. The metal bar is pulled by the magnetic field toward the bell and strikes it, causing the bell to ring, but the movement opened the circuit between points A and B, so current can't flow, the magnetic field collapses, and bar B falls back to its original position.

At that instant, the circuit is again complete, so the magnetic field rises and pulls bar B to hit the bell.

The process continues as long as the button is held in.

Figure 11.10 shows two reed relays. A *reed relay* is essentially two pieces of metal inside a tube. Reed relays can also come in different packages; for example, one looks like a small plastic box with connectors. They can be either normally open (NO, OFF) or normally closed (NC, ON). If the datasheet isn't available, the configuration is still easy to figure out. To test whether a reed relay is NO or NC, use a multimeter to measure the resistance. If the at-rest resistance is infinite, the relay is NO. If the resistance is very little, the relay is NC. In this regard, relays are like switches, and in fact they work the same by opening or closing a circuit. But in the case of a reed relay, the actuator is the magnetic field.

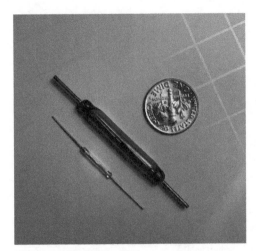

Figure 11.10: Reed relays

Reed relays in a circuit can be used in conjunction with a magnet bar in devices like alarm systems and automatic lights. Figure 11.11 shows one such alarm. When the door is opened, the magnetic field is moved away from the circuit, causing the relay to switch and an alarm to sound. This particular implementation of a magnetic relay circuit is connected via Wi-fi to a very loud audible alarm base in another room with its own Wi-fi connection. The alarm base is powered by *mains* power while the circuit on the door is battery powered. This type of circuit could be used anyplace where a door or window might open. It could also be connected to a myriad of circuits and used to turn anything off or on, including alarms, lights, fans, or even cameras.

Figure 11.11: Magnetic alarm

NOTE   The word *mains* is often used in diagrams to refer to a main AC power supply.

Relays

Other devices simply called *relays* are used in many different types of equipment. The main job of a relay is to use a smaller current to control a larger current, or several currents. A relay's coil might be used in conjunction with a controller like an Arduino or Raspberry Pi with an output of only 5V and a few mA of current, while its switch side can support up to hundreds of amps of current and thousands of volts. Either side of the relay may support DC or AC or both AC and DC. The key to finding this information is, once again, the product's datasheet.

A relay could also be used with a transistor circuit supplying the power to the coil, with a 555 timer sending pulses to turn it on or off, or even with something like a photovoltaic cell, water levels, temperature, motion sensor, or anything you want to use to turn a circuit on and off.

Parts of a Relay

I like the relay shown in Figure 11.12 because it has a clear case that lets you see what's going on inside. Many relays don't. This relay is a relatively small relay. According to its datasheet and the information printed on top of the housing, it can handle up to 5A and 240VAC (alternating current), or 5A and 30VDC (direct current) on the contact (switching) end. Its maximum wattage is 150.

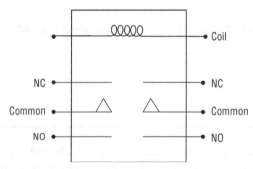

Figure 11.12: A relay

Notice in Figure 11.12 that one end of the relay is a coil and the other end has switches called *contacts*. The center bar is the switch's pole. On one side of the pole (closest to the coil) is an NC contact, and on pole's other side is an NO contact. This is repeated on the other side of the relay. This relay is a double-pole, double-throw switch, which can support up to four circuits, two of which will be on at any given point in time. The designation of NC or NO is known as the switch's *at-rest* position, that is, the position that the switch is in until the circuit is turned on.

Connect two alligator clips with leads to a 6V battery (one to the positive and one to the negative connector), then touch the opposite ends of the alligator clips to the two coil connectors, and you can observe the switching action of the relay. (Do this quickly and then stop. Be sure not to apply more voltage than your relay is able to safely use.) As soon as the relay is energized, a magnetic field builds and activates the switch. When the power is removed from the coil, the contacts return to their at-rest positions.

NOTE This demonstration relay uses 5V on the coil side and could be damaged if too much voltage, like that from a 9V battery, was applied. Be sure to check the datasheet for your relay's information.

Now that you know how a relay works, we're going to build a relay oscillator.

Try This: Building a Relay Oscillator

The main purpose of a relay is to use a smaller current to control a larger one. In this circuit, we'll do just that. For this project, you will need the following materials:

- Breadboard with two power input rails or a split power rail
- 9V battery with snap
- (3) AAA or AA batteries with battery holders and leads
- LED
- 1000uF capacitor
- 330-ohm resistor
- Small slider switch
- Relay (HLS = 14F3L or similar)
- Assorted jumper wires

Optional:

- Assorted capacitors

Use Figure 11.13 to assist in building your circuit. The following instructions reference the connection points on the demo breadboard. Yours may be slightly different.

Connecting the Coil

To connect the coil, follow these steps:

1. Place the relay over the center dip so that one side of the relay (one pole, one NC switch, one NO switch, and one coil connector) is on one side of the dip, and the other side of the relay is on the other side of the dip. The demo has the coil pins in 19f and 19g.

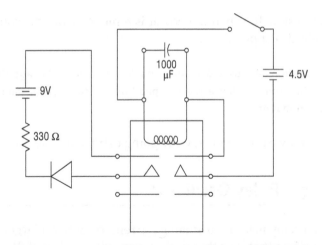

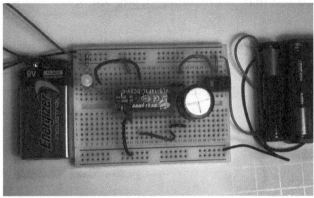

Figure 11.13: Relay oscillator circuit

2. Place two small jumpers between the coil and another junction far enough away to allow for a large capacitor (19j to 22j and 19d to 22d).

3. Place the capacitor's negative lead in 22h and the positive lead in 22f, connecting it between the two jumpers in the previous step, across the dip. This puts the capacitor in parallel with the coil.

4. Test the switch with an ohmmeter to make sure the two pins (center and one end) that you will be using are in the off (open) position. Resistance should be infinite or overload (OL), depending on your meter.

5. Place the switch so that the middle pin is in 29c and the off end pin is in 28c.

6. Jumper from the positive rail to 29a.

7. Jumper from 28a to 22a. This connects the switch to the coil and capacitor.

8. Jumper from the coil, 19k, to the NC switch, 13k.

9. Jumper from the common, 11h, to the negative power rail.

10. Place the AAA or AA batteries in the holder and connect them in series (i.e., the positive of one battery to negative of the next). Each battery is 1.5V, so series connections will cause the output to be 4.5V. The demo relay has a nominal voltage of 5V. The datasheet tells us that the maximum applied voltage is 6.5. A 9V battery on the coil side of the relay could damage it. When we combine batteries in this way, it is often called a *battery pack*.

11. Connect the battery pack's output leads to the positive and negative power rails in the appropriate spots (see Figure 11.12). Notice that the demo breadboard has split power (i.e., the red line is broken) between one end of the breadboard and the other. This project takes advantage of that feature, supplying 4.5V on one end and 9V on the other. If your breadboard doesn't have this split, you'll need to connect the power for the LED side of the circuit directly into the terminal row (details follow), or jumper from this breadboard to a second one.

12. Test the relay. Turn the switch on and with the switch activated, the relay should click. It will be clicking rapidly but should not be buzzing. Buzzing means that the relay is flipping on and off very fast. If it makes a buzzing sound, turn the switch off and check the connections of the capacitor. Make sure the negative side of the capacitor is toward the negative side of the circuit. The capacitor is used to slow down the switching action.

When the coil side of the relay is working, turn off the switch and continue on to configure the controlled (LED) side of the circuit.

Connecting the Controlled Circuit

To connect the controlled circuit, follow these steps:

1. Jumper between the power rail and 11a, connecting power to the common. Make sure the jumper is in the 9V side of the rail, not the 4.5V side. Alternatively, if the breadboard you're using doesn't have a split power rail, connect it directly to 11a when it's time to connect the power.

2. Jumper between the NC switch, 13b, and 5b.

3. Place the anode (positive lead) of an LED in 5e and the cathode (negative lead) in 3e.

4. Insert the 330-ohm resistor at 3a, using it to connect between the LED's negative lead and the negative power rail.

5. Connect the negative lead of the 9V battery to the same negative rail that the 330-ohm resistor is connected to.

6. Connect the positive lead of the 9V battery to the power rail that is jumpered to 11a or, if not using a breadboard with split power rails, connect it directly to the relay's common connector at 11a. The LED should be lit.

7. Flip the switch, and the LED should oscillate between on and off.

So, how does this work? Follow along on the schematic. The coil side of the relay is a complete circuit except for where it is broken by the open switch. When the switch is activated, the relay coil creates a magnetic field, and the capacitor is charged. The magnetic field activates the switch inside the relay. The closed circuit of the LED side is opened, and the LED is no longer lit. The switch inside the relay that is connected to the coil is also opened. Because the circuit is open, power is no longer supplied by the source, but for a time the transient response of the capacitor slows the collapsing of the magnetic field. When the capacitor is depleted, the common pole of the relay's switch falls back to its at rest position, completing the circuit, and the process starts over again. This is similar to the doorbell circuit described previously, and in fact a relay could be used to create a doorbell.

Try adjusting the transient response of this circuit by adjusting the capacitance. Remember that capacitors in parallel increase capacitance, while capacitors in series diminish it.

This project is merely a simple demonstration of a device with great capabilities. What if a motion sensor were connected to the coil side as a switch, and holiday lights or outdoor lights were connected to the switching side of the relay? What if a doorbell circuit on the controlled side were connected to a pressure-sensitive mat that acted as the switch on the coil side of a relay, letting you know when someone is standing on your welcome mat? What else could you turn on remotely or more safely? Personally, my table saw frightens me. I use it, but I don't like turning it on because the switch is behind the saw part. Some safeguards are in place, but it still makes me uncomfortable. It's poorly designed as far as I'm concerned. It also threw part of a 4 × 4 post at me one day, so it's lucky that we're still on speaking terms. If the switch bothers me enough, I could create a relay circuit with a simple switch and automatically turn on the table saw from the other side of the shop. A relay is a powerful tool in itself, and the possibilities for implementing it are limited only by your imagination.

Try This: Setting Up an Emergency Lighting System

Did you ever wonder how one of those emergency lighting systems work? You're about to find out. In fact, you can build one yourself! A relay makes it fairly simple. For this project, you will need the following materials:

▪ Breadboard with two power input rails or a split power rail (or two breadboards)

- ▪ (2) 9V batteries with snaps (or appropriate power for your relay)
- ▪ (2) LEDs
- ▪ 100-ohm resistor (R_1)
- ▪ 330-ohm resistor (R_2)
- ▪ Small slider switch (S_1)
- ▪ NO momentary push-button switch (S_2)
- ▪ Relay (HLS = 14F3L or similar)
- ▪ Assorted jumper wires

Use Figure 11.14 to assist in building your circuit. As in the previous note, the word mains refers to a main AC power supply. For this circuit, we're using DC, but the term mains is used in this project as a reminder that typically there would be a mains supply to the lights and most likely a DC backup, although a device called an *inverter* can be used to convert DC power to AC, depending on the needs of the user. In our case, both circuits are DC. The following instructions reference the connection points on the demo breadboard. Yours may be slightly different.

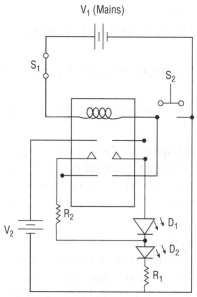
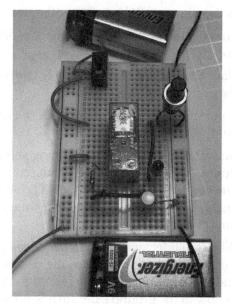

Figure 11.14: Emergency lighting circuit

Connecting the Mains Circuit

To connect the mains circuit, follow these steps:

1. Place the relay over the center dip so that one side of the relay (one pole, one NC switch, one NO switch, and one coil connector) is on one side of the dip, and the other side of the relay is on the other side of the dip. The demo has the coil pins in 19f and 19g.

2. Ensure that the pins you'll be using on the slider switch are in the off (open) position by testing with an ohmmeter.

3. Place the slider switch so that the middle pin is in 29c and the off end pin is in 28c.

4. Jumper from the positive rail to 29a.

5. Jumper from 28a to 19a. This connects the switch to the coil.

6. Jumper from the opposite side of the coil to the NO contact on the same side (19j to 9j).

7. Place the momentary switch (S_2) between the coil on that side (19k) and the negative rail.

8. Connect the positive lead of the first LED (D_1) to the same terminal row as the common (11k), and the other lead to 7k.

9. Connect the positive lead of the second LED (D_2) to 7j and the negative lead to 5j.

10. Place the 100-ohm resistor (R_1) between the negative lead of D_2 (5k) and the ground rail.

11. Connect the positive connector of the first 9V battery (V_1) to the positive rail that is connected to S_1.

12. Activate the slider switch. Nothing should happen. Follow the schematic from V_1's positive side, through the coil, and down to the relay's switches. The circuit is connected to the NO contact, so electricity can't flow to the LEDs. Above, notice that the momentary switch is also open, so electricity can't flow there.

13. With the slider switch (S_1) in the closed position, push and release S_2. The coil stays energized because the relay's switches have changed position, and the common is now connected to the NO switch, allowing electricity to flow to the ground side. Turn S_1 off and back on again. Notice that the lights stay off.

Connecting the Backup Circuit

The backup circuit is designed to be on whenever the mains power is off. The two circuits will share a common ground but will *not* share the same power rail.

1. Jumper from the other side of a split power rail to 13a, which is the terminal row of the NC switch on the opposite side of the LEDs. If there is no separate power rail, connect the power directly into 13a when the time comes (but not now).

2. Connect R_2 between 11c (the relay switch's common) and 7c.

3. Jumper from 7d to 7i, which should be the same terminal row where the two LEDs are connected.

4. Finally, connect the second 9V battery (V_2) so that its negative connector is in the same ground rail as the mains part of the circuit, and the positive connection is in the same power rail as the jumper to 13a, or if need be, connect the power directly into 13a.

NOTE V_1 and V_2 share a ground rail, but they must *not* share a power rail!

> D_2 should be lit. Follow the schematic around from the positive side of V_2 through the relay's switches, then R_2, and finally D_2. D_1 does not light because the function of a diode is to allow electricity to flow in only one direction, and there is not a complete path through D_2 to ground.
> When the relay is energized, its switches will change position. The circuit involving V_2 will be opened, so D_2 will no longer be receiving power from V_2. The circuit involving V_1 will be closed, and both D_1 and D_2 will light.

5. Ensure that S_1 is in the closed position.

6. Press S_2. Notice that both LEDs are lit.

7. Turn off S_1. D_2 is again getting its power from V_2. If you close S_1 again, D_1 will not light because the coil is not energized, and the switches are still in their at-rest positions.

We've barely scratched the surface of ways that magnetism can be used in electronic circuits, but now that you have some basic knowledge of how it works, let your imagination run wild and show the world what you can create with it!

Electricity's Changing Forms

"And the day came when the risk to remain tight in a bud was more painful than the risk it took to blossom."

—Anais Nin

If all electricity could do is flow around in conductors, it would be boring. We can't even see it doing that. Luckily, through the magic of circuits and the imagination of some incredibly smart people, we can make electricity move and shake things, light up and make sound, and so on. This is done, in large part, through the use of transducers. We've mentioned them before. A *transducer* is any device that changes between electrical current and some other form of energy, in either direction.

The sensors in IoT and other devices such as photoresistors, photodiodes, microphones, and generators (e.g., light, sound, and movement) are transducers that are used to input signals into an electrical circuit. Light bulbs, speakers, buzzers, and motors are some of the devices that convert electricity into some other form of energy and circuit output (e.g., light, sound, and movement). All of those will be examined, but first, let's make some noise.

Try This: Creating a Water Alarm

This is a fairly simple circuit that uses water as a conductor between two probes. When there is continuity between the two probes, a sound is emitted from the buzzer. For this project, you'll need the following materials:

- 9V battery and with snap
- Breadboard

- 100-ohm resistor
- 330-ohm resistor
- NPN transistor (BC 547)
- Buzzer, 1.5V–30V
- 22-gauge hookup wire, about 12 inches

Refer to Figure 12.1 as you build this circuit. If it looks familiar, it should. This circuit uses a transistor to allow a small current between the two probes to control a much larger current. For our circuit, we'll use two conductors with their ends stripped as the probes, but they could be almost anything conductive, provided the conductor is insulated except for the area that you want to use as a sensor.

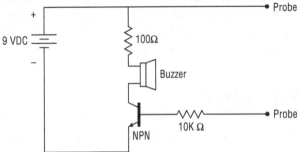

Figure 12.1: Water alarm

Follow these steps to build the circuit. Refer to Figure 12.1 to aid in placement of the components. The connections referenced here are for the sample board, so adjust the connections to match your board as necessary.

1. Place the transistor so that the base is in 9c, with the emitter in 8c and the collector in 10c. If you're using an NPN transistor different from the one specified in the materials list, be sure to check the datasheet for the pinout to ensure that it is inserted correctly.

2. Place a 100-ohm resistor with one lead in 13d and the other in 15d.

3. Place a 330-ohm resistor with one lead in 4d and the other in 8d (in the same terminal row as the transistor's emitter).

4. Insert the positive lead of the buzzer into 13a.

5. Insert the negative lead of the buzzer into 10a (in the same terminal row as the transistor's collector).

6. Insert one lead of a long jumper wire (probe) into 15a.

7. Insert one lead of a long jumper wire (probe) into 9a (in the same terminal row as the transistor's base).

8. All that's left is the power source for the buzzer. Insert the positive lead of the 9V battery in 15e (the same terminal row as the first probe).

9. Insert the negative lead in 4e (the same terminal row as the end of the 330-ohm resistor).

The buzzer is one that will activate at a very low voltage (1.5V) and emit a sound, although a higher voltage will produce a louder sound. It's best to use a buzzer that is capable of voltages slightly higher than the input voltage of the circuit. Although 30V is excessive for this circuit, buzzers in that range, like the one purchased for this project, are inexpensive and easy to find. They are often called mini buzzers, and at the time of this writing, a pack of five from a major online supplier runs about $10.

How does this circuit work? The resistors are there to limit the current on the circuit. The yellow (lighter) probe shown in the picture is connected to the collector of the NPN transistor. The blue (darker) probe is connected to the base. When the two probes are connected in some way—whether they both touch your finger, the same body of water within close proximity to each other, or your tongue (yes, I tried it; it tingles)—the connection of the two probes completes the circuit. The connection goes from the (–) of the source through the emitter and base of the transistor and back to the positive side of the source, causing electricity to flow. This small current on the base, in turn, "opens" the channel between the emitter and collector and allows a larger current to flow through them and the connected buzzer. (Refer to Chapter 9, "Transistors," for more detailed information.)

That is how the transistor works, but this chapter is about transducers, so the bigger question is, how does the buzzer turn electrical energy into waves of sound energy? I took a buzzer apart to show you.

Refer to Figure 12.2. Inside the buzzer housing is a circuit board and a piezo element. Piezoelectric materials are a group of materials that can transmit an electrical charge when pressure is applied to them. They also work the opposite way and exhibit mechanical movement when an electrical charge is applied *to* them. The action of piezo materials can be precisely calculated, and piezo materials are used in industries from aerospace to medical equipment to devices you use every day, such as inkjet printers and liquid crystal display (LCD) panels.

Figure 12.2: Inside a buzzer

The piezo element in Figure 12.2 consists of a ceramic disc surrounded by a metal vibration disc. When electricity is applied to the ceramic disc, it will contract or expand, causing the metal disc to move. This movement, in turn, creates air pressure and causes a sound wave as it undulates back and forth.

The circuit board holds an oscillator circuit, which sends pulses of electricity at a particular frequency. The frequency of this oscillation determines the pitch of the sound wave.

This is just one type of transducer; however, there are many more.

Common Transducers

As previously mentioned, a transducer is any device that converts between electricity and some other form of energy. You already have learned that piezo materials can be used to create sound, but how do other speakers work? What about microphones? How do they take the sound of our voice and turn it into electricity that can be transmitted across a circuit? Let's take a deeper dive into those and other transducers.

Speakers and Microphones

Speakers are like the piezo cell in that they both must create air waves in some way for us to be able to hear sounds. They differ, however, in the method used to achieve that. At a minimum, a speaker has a coil that is used to create an electromagnet, a permanent magnet, and a cone (diaphragm) to move the air, and of course it is connected to a circuit. See Figure 12.3.

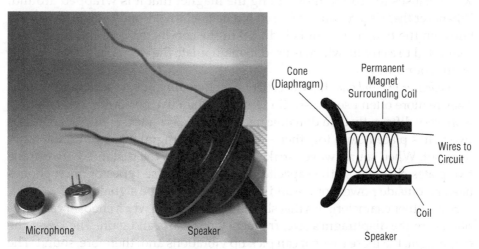

Figure 12.3: Speakers and microphones

The coil will be attached in some way to the cone of the speaker, which is usually made of something flexible like paper. The circuit passes electrical pulses to the coil. The magnetic field created by the coil is attracted to, or repelled from, the permanent magnet. This force causes the coil to create physical movement, which pulls and pushes on the diaphragm. Think of the cone, or diaphragm, of the speaker as being like the diaphragm beneath your lungs that moves to bring air in and to force air out. As a speaker's diaphragm moves, air is pushed, and the moving air creates waves of sound. The frequency of the wave determines the pitch of the sound. A faster frequency will result in a higher sound, like a bird chirping. A slower frequency will result in a deeper sound, like a bullfrog croaking.

Because the higher frequencies must move the cone faster, speakers that create higher sounds will often be smaller (i.e., have less mass to move) and are called *tweeters*. Speakers that make the lower sound are called *woofers* and are usually larger. Of course, there are midrange speakers, too.

The circuit connected to each speaker type is designed to let a certain range of frequencies pass while absorbing others. These are high- or low-bandpass filters, which will be discussed in more detail in the later chapters that address alternating current.

Microphones work similarly to speakers, but in reverse. The air pressure created by waves of sound move a membrane that translates into electrical pulses. There are many different types of microphones, each with characteristics that make them better suited to one situation than another. Here are just a couple.

Dynamic microphones are popular, rugged, and versatile. These mics are often used for live stage performances. A dynamic microphone uses a coil and magnet like a speaker does, but it operates in reverse. Air waves push on a diaphragm, which causes the coil to move along the magnet that it is wrapped around. Whenever there is a conductor (the coil), a magnetic field, and relative movement between the two, a current is induced in the conductor. The conductors are connected to a circuit, which is used to manipulate the electrical pulses created by the sound waves.

Condenser microphones have excellent sound quality but are a bit more fragile. They're more often used in studio recording than live performances, and they work very differently than dynamic microphones. A condenser microphone uses two plates placed close together—one (the diaphragm) moves, and the other does not. When sound waves strike the diaphragm, the distance between the two plates changes, so the capacitance changes. These types of microphones need an outside power source and circuitry to boost the signal.

Some other characteristics that should be considered when choosing a microphone are the diaphragm's size, frequency response, and gain. The larger the diaphragm, the more easily it can pick up vibrations and, therefore, sound. The material it's made of affects its weight and sensitivity. Frequency response refers to the range of frequencies that the microphone can pick up. Is it better at picking up high frequencies or low ones? Finally, gain is a measure of how much stronger the output is than the input. Too great a gain can cause distortion, but too little might drop weaker sound waves and miss the nuances of voice and music.

Light

Light can be both input to a circuit and output from a circuit. First, we will look at the devices that use light to control a circuit; then we'll look at how several different types of lamps work.

Light-Controlled Devices

The combination of light and electronics in theory and actual use is known as *optoelectronics*. Light can be produced from electricity, but light is also used in electronics for a number of purposes. The light can be within the visible spectrum, but nonvisible wavelengths on the electromagnetic spectrum, such as ultraviolet and infrared, are also used in optoelectronics.

Optoelectronics are perhaps more easily recognized by the prefix *photo*. Whenever the name of a device is preceded by *photo*, the device uses light in some way. Photodiodes and phototransistors conduct current based on how much or how little light is reaching the devices. Photoresistors, also called light-dependent resistors (LDRs), have more or less resistance based on the light that reaches them. Photovoltaic (PV) cells convert light energy into electrical current. (PV cells are discussed more in Chapter 15, "Sources of Electricity.")

A matched pair consisting of a photodiode and either a laser or LED can be used to transmit and receive information over fiber-optic lines, which are the backbone of many of our modern communication systems. Photo devices are also used in alarm circuits, night lights, night vision goggles, camera flash circuits, and any device that needs to base its function on the light it receives.

A *photodiode* uses the now familiar PN junction. For an extremely simplified explanation, consider the following: when a photon of light passes into the photodiode, it strikes atoms in the p-type material, knocking electrons out of their orbit and causing free electrons and an excess of holes in the p-type material. Some of those free electrons will cross the junction and cause an excess of electrons in the n-type material. A potential difference exists between the p-type and n-type material. So, just like the poles of a capacitor, if the cathode and anode of the diode are connected via a circuit, electrons will flow from the more negative cathode to the anode. The more photons of light that reach the p-type material, the more electrons that will be set free and holes created, causing a greater potential difference. In the case of a photodiode, typically the light received is directly proportional to the amount of current generated. Photodiodes often look like a small canister with a little "window" in the top to allow light in.

Phototransistors may look like an LED (or like the photodiode described earlier), but they are a completely different device. Typically, a bipolar junction transistor will have three leads: an emitter, a base, and a collector. In the case of a phototransistor, the lead on the base is omitted and is replaced by light striking the phototransistor. They are often made to look like an LED, because a large p-type area is covered with a clear epoxy, leading to the LED appearance.

On a typical NPN transistor, a voltage applied to the base causes a current to flow between the base and emitter and "opens" the connection between the collector and emitter, allowing a much larger current to flow there. In the case of a phototransistor, instead of a voltage applied to the base, light reacts with the p-material in the base, causing electrons to be freed and move to the emitter region (i.e., current flows), opening the connection between the collector and emitter, and thereby controlling a larger current. The more light, the more the "valve" between the collector and emitter opens. The gain on a phototransistor may be much more than the gain on a photodiode.

Phototransistors may be used in applications like manufacturing, where counting or timing is needed (e.g., as boxes on a conveyor pass through a light beam, breaking the transmission from the light source to the phototransistor on the other side). The time the light is on and off can be measured, and the speed can be controlled by circuits, or the circuit could simply count the number of times the light beam is broken. Phototransistors are also used in alarm systems, and they can even be used to isolate circuits yet still pass the frequency from one circuit to another.

Photodiodes and phototransistors come in several different physical package configurations. Some will have more leads than the typical two that were discussed here. One of the biggest differences between photodiodes and phototransistors is the ability of the phototransistor to control a larger current.

When choosing a phototransistor or photodiode, the datasheet needs to be referenced not only for the maximum voltage and current but also for the light characteristics that control the device. See Figure 12.4 for typical light-based circuit control devices.

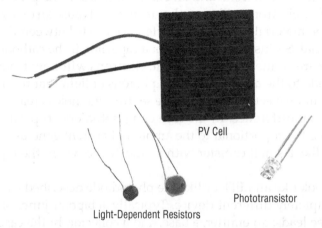

Figure 12.4: Light-based circuit control devices

A typical *light-dependent resistor* (LDR) looks like a small plastic disc with a squiggle line across the top. They usually have two leads. An LDR is simply a variation on the same theme as the phototransistors and photodiodes. LDRs are made of doped semiconductor material such that when light strikes the material, electrons are dislodged from their atoms and current flows. The more light, the less resistance there will be.

An LDR may have a resistance of several megaohms when in the dark, and only a few ohms when in the light. The actual resistance varies depending on the LDR, and again the datasheet needs to be consulted. Depending on what material they are made of, the LDR may react to room light, daylight, or infrared light. LDRs are used in devices such as patio garden lights, streetlights, and night lights, but they can also be used to turn a circuit on when light is detected, as well as when darkness is detected. The exact use is dependent on the circuit the LDR is attached to.

Light Output Devices

An entire book could be devoted to the different light output devices available. Here are some of the most prevalent.

Incandescent Lights

A simple incandescent light bulb is a transducer that converts electrical energy into light and heat energy. The incandescent bulb has a tungsten thread, called a *filament*, which is heated to a point that it glows. Incandescent bulbs are inefficient because so much of the energy is converted to heat rather than light. Tungsten is chosen as the filament because it has a high melting point. During the normal course of use, the tungsten will vaporize and be deposited on the glass of the bulb. The familiar bulb is a vacuum tube that has been depleted of oxygen, because otherwise the oxygen would degrade the tungsten much more quickly and the bulb wouldn't last long.

A halogen bulb is also a type of incandescent lamp. It has a tungsten filament like a regular incandescent lamp, but the bulb is filled with halogen and inert gases. The tungsten in these lamps vaporizes as it does in an incandescent lamp. However, instead of depositing on the bulb, it reacts with the halogen gas and is deposited back onto the filament. The envelope containing the filament in a halogen lamp is stronger than the glass of an incandescent light so it can withstand greater pressure and heat. The advantages of halogen bulbs are that they last longer and can produce more light because they are able to operate at a higher temperature. See Figure 12.5 for a sampling of light bulbs.

Halogen   Mercury Vapor   Compact Fluorescent

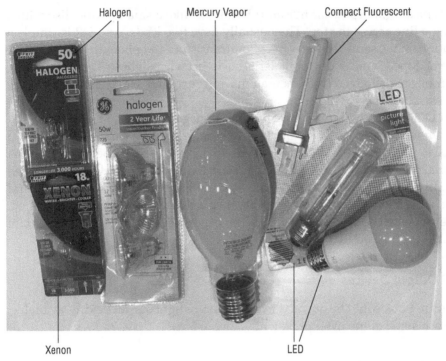

Xenon   LED

Figure 12.5: Light bulbs

Discharge Lamps

Lamps that use either glowing gases, vapors, or phosphors are called *discharge lamps*.

Neon Lights

The ubiquitous neon light that can be seen in "Open" signs and almost every roadside eatery and pub is a type of discharge lamp. Neon lights are named after the gas they are filled with, although since their inception they have been filled with other gases to produce different colors, which out of custom are still called *neon lights*. They require a very high voltage, so a step-up transformer is used to generate voltages of 10,000V or more. Each end of a neon light tube has an electrode. When sufficient voltage is applied, the atoms of the gas become excited and electrons jump to a higher level. As they fall back to a lower level, a photon of light is produced. Argon and helium are often used to produce different colors.

Fluorescent Lamps

Fluorescent lamps work in much the same way as neon lights, requiring a very high voltage; however, they contain mercury that is vaporized inside the tube. Electrons flow along the mercury vapor from the electrode on one end to the electrode on the other end. The process produces UV light, which is invisible to the human eye, so fluorescent tubes are coated inside with phosphorous material that fluoresces (glows) when hit by the UV light.

Compact fluorescent lamps (CFLs) are used to replace incandescent light bulbs. CFLs are much more efficient than incandescent bulbs because they produce more lumens of light per watt and release less energy as heat. Some are designed to screw into the same socket as incandescent bulbs, while others have different configurations. A quick look at a hardware store website will show several options. A CFL that produces lumens equivalent to a 100W incandescent light bulb would typically use only about 23W of power.

No discussion of light output devices would be complete without mentioning the familiar LED. LEDs are a type of diode that, when forward biased with the right voltage, will emit light of a particular wavelength. This happens when an excited electron recombines with an atom and loses some of its power, which is emitted as a photon. Different materials will emit different colors within the visible spectrum or invisible wavelengths, such as infrared and the like. The color emitted varies, based on the amount of energy released. LEDs are packaged in many different ways— there are the familiar through-hole LEDs, but also surface mount packages that are about the size of a grain of rice like the LED attached to pin 13 of the Arduino board, LED stickers, and mounted in every conceivable package of "light bulb" with various connectors, from the ones compatible with the incandescent bulb fixtures in your home, to car headlights, floodlight bulbs, simply soldered on, and everything in between.

Often LEDs are packaged together to create a brighter package like the Adirondack camp light in Chapter 5, "Dim the Lights." They also come in bar and seven-segment packages like the project in Chapter 8, "Diodes: The One-Way Street Sign."

Others

Mercury vapor lamps can often be seen on outdoor poles and are used on farms, in streetlights, in parking lots, and in stadiums. They produce a great deal of light and are long-lasting, although it takes a few minutes to get them glowing after the power is turned on. They typically contain mercury and argon. A starting electrode sends power through the argon gas to a main electrode. As the electricity arcs through the argon gas, heat is produced. Eventually the mercury becomes vaporized from the heat and arcing across the mercury vapor produces a bright light.

Strobe lamps are used in places like airport runways and emergency lights. They may contain a gas, such as xenon, and as in the mercury vapor lamp, the bright light is produced by electricity arcing through a tube. Strobe lamps use a much higher voltage and can flash several times a second. More recently, LED lamps connected to circuits pulsating at a specified frequency have been used as strobe lamps. The effect of a strobe lamp can also be created by covering a portion of a light with a rotating cover.

A glow lamp is a low-voltage version of a neon lamp. They are typically very small.

Laser Light

Laser stands for "light amplification by stimulated emission of radiation." Laser light can be both input to and output from electronic circuits. Lasers have many uses in modern life. They are used in medical equipment for precise cutting and cauterizing. They can be found on construction sites to cut metal and to level a 2 × 4 when building a wall. Lasers can also be found in the optical drives of computers, and they make amusing cat toys. (See Figure 12.6.) On the flip side, lasers are used in data communication across fiber-optic networks and in transferring data from a digital format to the photosensitive drum of a laser printer. Obviously, there are many different types of lasers for many different uses. While laser diodes and the safety precautions were discussed in Chapter 8, that information bears repeating here.

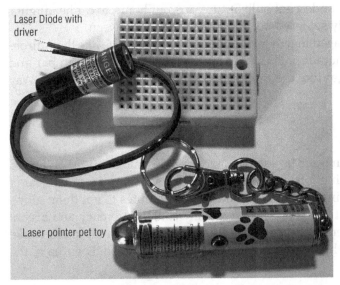

Laser Diode with driver

Laser pointer pet toy

Figure 12.6: Laser pointer

WARNING   Lasers can be dangerous! Never point a laser at a car, airplane, another person, or pet. Even reflected laser light can cause permanent and irreversible eye damage or blindness. Lasers can also cut or burn. Proper eye protection *must* be worn when working with dangerous classes of lasers.

Make sure you know what type of laser you have and ensure that you have proper protective equipment before you power it on. This is one area where scavenging parts from something that no longer works may not be a good idea. Look for markings on the laser to indicate its class, and refer to the manufacturer or the OSHA website for instructions on how to work with that particular type of laser. At the time of this writing, that website is www.osha.gov/SLTC/laserhazards/.

As in some other light sources, the light occurs when excited electrons lose their energy and drop back to a lower atomic shell. The difference with laser lights is that the wavelength of the photons is the same, so a laser is a specific color. A laser light is unidirectional, rather like an LED output and not diffused like the sun, so it can travel long distances without *attenuation*, which is loss of signal strength.

Other Transducers

Motors and generators were discussed in Chapter 11, "The Magic of Magnetism." Both motors and generators are transducers. However, a generator takes mechanical movement of some sort and turns it into electrical current, while a motor converts electrical energy into mechanical movement. We'll work more with motors in Chapter 14, "Pulse Width Modulation," and with generators in Chapter 15, "Sources of Electricity."

Certain fiber optics can be used as transducers to sense pressure changes in buildings and bridges, thereby monitoring their structural integrity.

A thermocouple could be considered a transducer. In a thermocouple, two dissimilar metals are connected together. An increase in ambient heat will induce an increased current flow in the metals. Thermocouples are used in ovens and other devices where temperature control is important. There are of course many other transducers; we have covered only a few.

Finally, a note about transducers versus sensors. While they work closely together and the terms are often used interchangeably, they are *not* the same. A sensor measures changes in something physical—more or less pressure, more or less light, more or less temperature, or more or less moisture. A transducer changes that measurement into something that we can interpret, such as a voltage, or conversely, from voltage to light or heat.

Now, it's time for some fun with transducers.

Try This: Creating a Night-Light Circuit

This simple little circuit is a staple in the electronics world. Team it with a relay, and anything can be made to turn on when the sun goes down. For this project, you need the following materials:

- Breadboard
- 9V battery and snap
- Photoresistor 1k–15k ohms
- 330-ohm resistor
- 100k-ohm resistor
- NPN transistor (BC 547B)
- Small jumper
- LED, any color

First, use a multimeter to determine the value of the photoresistor when it is in darkness and light. The one used in this circuit measured 3.3k ohms in the light and 15.5k ohms in darkness. If the resistance is too high in the light or too low in the dark, the circuit may not work.

The transistor used here is a BC 547B, but a general-purpose NPN transistor like a 2N3904 will work just as well. If using an alternate transistor, be sure to check the pinout and adjust accordingly. Use Figure 12.7 to assist in building your circuit.

The following instructions are for the demo breadboard. Your board may be numbered differently, so adjust the connections as needed.

1. Insert the NPN transistor so that the emitter is in 10h, the base in 9h, and the collector in 8h.

2. Insert the LED with the negative lead in 8i (same row as the collector) and the positive lead in 7i.

3. Insert the 330-ohm resistor with one end in 7j and the other in 5j. (It will connect the LED to the positive side of the source.)

4. Insert the 100k-ohm resistor with one end in 9g (same row as the base) and the other in 5g (same row as the 330-ohm resistor).

5. Insert the photodiode with one lead in 12j and the other in 9j (same row as the base).

6. Jumper between 10i (same row as emitter) and 12i (same row as the photodiode).

7. Connect the positive side of the source to 5f (the row with both resistors) and the negative side of the source to 12f (the row with the jumper and the photodiode).

8. To test the circuit, place your finger over the receptor of the LDR and the LED should light.

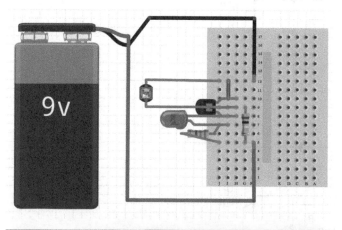

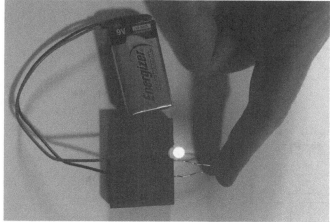

Figure 12.7: Night-light circuit

To understand how this circuit works, review transistors. The transistor must be forward-biased, which means when a more positive voltage is applied to the base of an NPN transistor than exists on the emitter, a small current will flow between the base and the emitter. This "turns on" the transistor and allows a much larger current to flow between the collector and the emitter. See Figure 12.8.

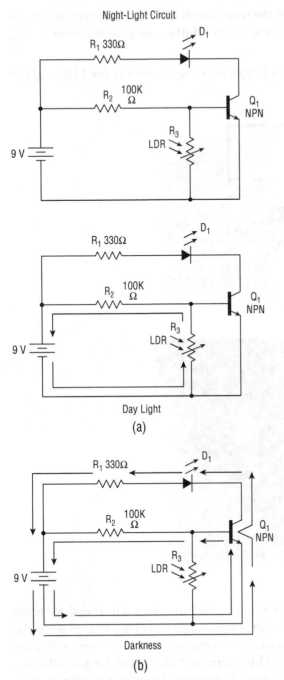

Figure 12.8: Electron flow in the night-light

When the LDR is in daylight, the resistance is low, so electrons will take the path through the LDR and the 100k-ohm resistor back to the source. (See Figure 12.8a.)

When the LDR is in darkness, its resistance is high, and electrons will take the easier path through the emitter and the base of the transistor back to the source. (See Figure 12.8b.) This "turns on" the transistor, so a greater current can flow through the emitter, collector, LED, and 330-ohm resistor, and back to the source.

The breadboard used in the demo circuit has a peel-and-stick back, so it could easily be put almost anywhere, but it might be a great circuit to mount more permanently on a perfboard.

Now, get those creative juices flowing. Imagine. Long jumper wires could be connected to either the leads for the LDR or the LED, or both, so darkness in one area could turn something on in another area, and connecting a relay instead of an LED would let you turn anything on or off based on light. The NPN transistor could be replaced with a PNP transistor, which would essentially work the opposite way—it would be on in the daylight and off in the dark.

> **NOTE**   There is some, although small, resistance even in a good conductor. So technically, there is a limit to the length a wire can be and still work well. Keep this in mind if you're planning to run wires a long distance. Remember Ohm's law.

Try This: Creating an Arduino Laser Security System

The physical connections for this circuit are relatively simple. Parts of it may look familiar because they have been created in previous projects. The programming of the Arduino is a bit more complicated but still fun!

This project explains how to make the laser alarm system work, and at the end some options for implementing it into an environment will be discussed.

Building the Circuit

First, we'll build the laser circuit. A laser diode by itself will not work properly, because it needs a driver circuit. The driver circuit ensures that the current reaching the diode is appropriate—not too much nor too little. For this project, we're using a laser diode that has an integrated driver, which makes constructing our project that much easier. All we need to add to the laser circuit is a power source and a way to turn it off and on. For the power source, we'll use a battery pack of three batteries in series (4.5V), because this laser uses a power source between 3–5 VDC. The Arduino Uno provides either 3.3V or 5V, but because the laser and alarm circuit will be separated by enough space for something to trigger it, using the Arduino board for the laser circuit would not make sense. The laser side of the circuit is so simple, in fact, that using the Arduino is not necessary. In the absence of a laser diode circuit, a simple laser pointer toy can be made to work.

WARNING Be sure to read the warning label on the laser or laser pointer that you use for this project! Use appropriate caution when handling a laser. Never point it at a person or animal's eyes or at an airplane or car.

Next, let's look at the sensor/alarm circuit. The sensor circuit uses a voltage divider that incorporates a photoresistor (LDR). The voltage across the voltage divider will be split proportionately between the fixed resistor and the LDR. When the LDR is in darkness, its resistance and voltage drop go up, and the voltage reading between the LDR and the fixed resistor go down. At that point, the sketch (i.e., program) in the Arduino will turn on the buzzer, which will beep until the laser light is restored.

For this project, you need the following materials.

For the laser circuit:

■ Mini breadboard

■ (3) AA or AAA batteries and holders

■ Laser diode with integrated driver, Adafruit 1054 or similar

■ Slider switch or push-button (not momentary)

■ Jumper wire

For the sensor/alarm circuit:

■ Arduino Uno

■ USB/Arduino cable

■ 9V battery with Arduino connector

■ Breadboard

■ Jumpers, assorted lengths

■ LDR (photoresistor) 1k–15k

■ Resistor, value to be determined, approximately 1k ohms

■ (2) LEDs, one green, one blue

■ Buzzer 1.5–30VDC or similar

Optional:

■ Double-stick tape or mounting tape

■ Extra lengths of hookup wire

■ Project box(s)

To assist with component placement, check out Figure 12.9. Use the following instructions for the laser portion of the circuit. Your breadboard may be different than the demo breadboard, so adjust accordingly.

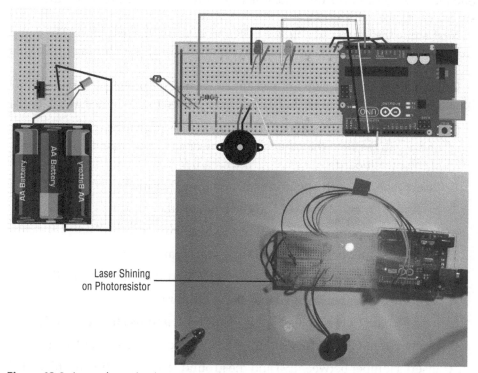

Figure 12.9: Laser alarm circuit

1. Insert the red lead of the laser in 1a.

2. Insert the black lead of the laser in 4a.

3. Insert the slider switch so the center pole is in 5e and an end connector is in 4e.

4. Ensure that the batteries are connected in series. The voltage rise across the battery pack should be 4.5V.

5. Connect the positive lead of the battery pack to pin 1e and the negative lead of the battery pack to 5c. That's it!

Flip the switch to ensure that the circuit works (hold your hand or a paper in front of the laser; don't look directly at it), and turn the circuit off until it is needed again.

Now it's time for the alarm part of the circuit. Before you begin, measure the resistance across the LDR when it is in the brightest ambient light it will have, with the laser. To make this easier, try wrapping one lead of the LDR around one meter probe and the other LDR lead around the other probe. Jot the measured resistance value down.

Next, shine the laser light onto the LDR and measure the resistance. The demo LDR measured about 1,000 ohms in daylight with the laser shining on it. This is its minimum resistance.

Refer again to Figure 12.9 for assistance with component placement. Your breadboard may be different than the demo breadboard, so adjust accordingly.

1. Jumper across the breadboard from one positive rail to the other and from one negative rail to the other so that power can be accessed on either side of the breadboard.

2. Insert the LDR with one lead in 3i and the other in 5i.

3. Insert a jumper between 3j and the power rail.

4. Insert the 1k-ohm resistor between 5h and 9h.

5. Insert a jumper between 9j and the ground rail.

6. Connect the red lead of the buzzer in 15h and the black lead in 17h.

7. Insert a jumper between 17j and the ground rail.

8. Insert a blue LED with its long (+) lead in 17c and its short (-) lead in 19c.

9. Insert a green LED with its long (+) lead in 24c and its short (-) lead in 26c.

10. Insert a jumper between 19a and the ground rail, connecting the blue LED to ground.

11. Insert a jumper between 26a and the ground rail, connecting the green LED to ground.

12. Connect a red jumper wire between the 5V pin in the power area of the Arduino and the breadboard's power rail. If the power rail has a split between one and the other, ensure that the power is connected to the power rail with the components.

13. Connect a black jumper wire between the GND pin on the Arduino and the ground rail of the breadboard, again making sure the components will connect to it.

14. Connect a jumper between Analog I/O pin A1 and breadboard pin 5f. It will be in the same terminal row that connects the LDR and the 1k-ohm resistor.

15. Connect a jumper wire from digital I/O pin 7 to breadboard pin 15f. It will be in the same terminal row as the red lead of the buzzer.

16. Connect a jumper wire from digital I/O pin 5 to breadboard pin 24a. It will be in the same terminal row as the positive lead of the green LED.

17. Connect a jumper wire from digital I/O pin 3 to breadboard pin 17a. It will be in the same terminal row as the positive lead of the blue LED.

18. Use the USB cable to connect the Arduino to a PC with the Arduino IDE installed.

19. Open the IDE and enter the following code. It will be explained after the end of the code. You may omit the lines with // to save time, but you may want them there later to remember what each line does.

```
// sets the pin reading the input value to analog pin 1, and names
it readIt
// initializes the variable inputValue and sets the value to 0
// sets pin 5 to be the on indicator and names it onPin
// sets pin 3 to be the off indicator and names it offPin
// sets pin 7 to be the buzzer and names it buzzerPin
// defines the on/off threshold value and names it trigger

int readIt=A1;
int inputValue=0;
int onPin=5;
int offPin=3;
int buzzerPin=7;
const int trigger = 300;

void setup() {
 Serial.begin(9600); // communicate on the serial port

  // sets the pins to input or output
 pinMode(readIt,INPUT);
 pinMode(onPin,OUTPUT);
 pinMode(buzzerPin,OUTPUT);

  // initializes the pin values
 digitalWrite(onPin,LOW);
 digitalWrite(offPin,LOW);
 digitalWrite(buzzerPin,LOW);

}

void loop() {
 digitalWrite(onPin,LOW); //resets the onPin to zero
 digitalWrite(offPin,LOW); //resets the offPin to zero

  inputValue=analogRead(readIt); //reads the value on the input pin
 Serial.println(inputValue); //prints the value on the serial
 monitor
```

```
// if the inputValue is higher than the trigger value, the onPin
lights up
// otherwise the offPin lights and the buzzer sounds

if(inputValue >= trigger) {digitalWrite(onPin, HIGH);}
else if(inputValue < trigger) {digitalWrite(offPin, HIGH);
digitalWrite(buzzerPin, HIGH);

// the next 3 lines cause a 1/2 second delay while the buzzer is
on, then turns it off
// and another 1/2 second delay before reevaluating the circuit
delay(500);
digitalWrite(buzzerPin,LOW);
delay(500);}

else {} //nothing
delay(500); // Without this delay the green LED flashes
// on and off so fast that it appears to be dim.

}
```

20. Click the check mark to verify the code. Make corrections as necessary and check again.

21. Click Tools and then Serial Monitor to open the Serial Monitor.

22. Again from the Tools menu, verify that Arduino Uno is selected next to Board.

23. Finally, from the Tools menu, click Port to ensure that the port is selected.

24. Click the upload arrow to upload the sketch to the Arduino board. The buzzer should start sounding within a few seconds.

If the buzzer doesn't start beeping, carefully check the circuit connections and the code entries. The program may also need an adjustment as described next.

How the Voltage Divider Works

As previously mentioned, the resistor R_1 and the LDR (R_2) act as a voltage divider. In theory, if the resistance of the LDR with the laser light shining on it measured 1,000 ohms and a 1,000-ohm resistor were used, then the voltage across each when the laser is reaching the LDR would be approximately 2.5V. The LDR or the resistor may not be quite perfect due to the resistor tolerance, and the presence or absence of ambient light might affect the reading of the LDR. In the darkness with the laser on, the LDR resistance is less than 1k ohms. When the laser is interrupted and there is no ambient light (darkness), the demo LDR's resistance spikes to 15k ohms. The result is about 0.3V drop across the resistor and 4.7V drop across the LDR. In daylight without the laser, the resistance of the LDR measured about 2.2k ohms, so in daylight if the laser is interrupted, the voltage drop across the LDR would be about 3.44V ((5V/(1+2.2))*2.2), while the voltage drop across R_1 would be 1.56V.

Remember that the voltage measured on an analog in pin will be read by the Arduino as a binary value between 0–1023, with 1023 being equivalent to 5VDC. The voltage of 2.5V would equate to a value of 512, and 4.7V would equate to a value of 962 ((1023/5)*4.7). When the laser is off, the resistance of the LDR is much higher, so the majority of the voltage drop is also there. That means that the reading taken by pin A1 will be much lower when the laser is off. Figure 12.10 may help explain this concept.

	Resistance		Voltage drop			Analog In
More light	R1	LDR	R1	LDR	Total	Value
	1000	500	3.33	1.67	5	682
	1000	1000	2.50	2.50	5	512
	1000	1500	2.00	3.00	5	409
	1000	2200	1.56	3.44	5	320
Less light	1000	15000	0.31	4.69	5	64

Figure 12.10: Calculating analog in values

The voltage drop across each resistor is the value of that resistor divided by the sum of the resistors, multiplied by the 5V input value. The total voltage drop across the two resistors will always equal the voltage going into the resistor pair.

The analog in value is (1023/5)*VR_1, and 1023/5 gives the value per volt times the voltage drop of R_1, which gives R_1's input value on pin A1.

Take the following steps to figure out the maximum value of the analog input pin:

1. Measure the resistance of the LDR (R_2) in ambient daylight.

2. Measure the value of R_1.

3. Now calculate, (1023* R_1) / (R_1+R_2) = maximum analog in value of R_1 – laser on.

Examine Figure 12.10. Find the cell where the resistance value of the LDR is 2,200. If in the daylight, with the laser off, the resistance of the LDR is 2,200 ohms, it will always be lower with the laser light shining on it. At 2,200 ohms of resistance on the LDR, the analog in value of R_1 is 320. Moving up the chart to an LDR resistance of 1,500, the analog in value is 409. Moving down the chart where the LDR resistance is 15,000, which it would be in darkness without the laser, the analog in value of R_1 is 64. With the laser off, the analog in value will always be lower because R_1's voltage drop is lower when the resistance on the LDR goes up.

Armed with this information, if I set the alarm to go off when the value read on the analog pin is < = 320, then the alarm should stay off as long as the laser is in contact with the photoresistor (LDR). This could change, depending on the ambient light in the area where the LDR is, so this value may need to be tweaked a bit for your specific circumstance.

In the sketch for this circuit, the variable called `trigger` is the threshold between when the buzzer should sound and when it should not. If the circuit is working properly, the Serial Monitor is displaying the reading between the LDR and the fixed resistor. You may want to disconnect one lead of the buzzer to test this value—the blue light will be on when the buzzer would be, so you can still tell if it's working without the sound.

Try shining the laser directly on the LDR and observe the measurement on the Serial Monitor. Then, observe the measurement without the laser light. If possible, observe the measurement in the lighting conditions that would exist where the laser alarm will be.

Open the sketch in the IDE, and change the value in the line that says `const int trigger = 300;` to a value that works for your circuit. It does not need to be changed anywhere else, because elsewhere in the sketch the value is referred to by its variable name. If you make a change, remember to verify and upload the sketch to your Arduino board.

Once the sketch is working, place the laser circuit and alarm circuit on opposite sides of the area that you want to monitor, such as a doorway. It's important to precisely line up the laser pointer with the LDR. When an object, person, or pet is blocking the light beam from the laser, the alarm will sound, and the light will change from green to red.

Other Considerations

If this circuit were being used outside, it's important to have a project box or other enclosure to protect the circuit. In that case, you may want to permanently mount the circuit on a perfboard.

Ambient light may make a difference, so you might need to try the circuit in its natural habitat and tweak the trigger value.

As this is a beginner book, the circuit is rather basic. However, much can be done to make it more sophisticated. Here are some ideas: it would be an excellent circuit to control via Wi-Fi. The addition of a keypad to change the trigger value and to turn the alarm off and on would also be good. The sketch could be changed so that once the beam has been broken, the buzzer will continue to beep until a code is entered into the keypad, even if the beam is restored. An LCD screen could also be added for visual output. With a relay attached, the circuit could be used to control a much louder alarm or turn on lights. The circuit could also be attached to a camera and cause it to turn on and record if the alarm is tripped. Alternately, an audio could play telling someone they are not allowed in this area when the alarm is tripped.

I'm sure your creative mind could think of even more possibilities. Enjoy.

CHAPTER 13

Integrated Circuits and Digital Logic

"Logic is the beginning of wisdom. . .not the end."
—Mr. Spock, *Star Trek VI: The Undiscovered Country*

Pop open any computer or logic device and you'll most likely see not one but several integrated circuits. These devices, sometimes called *computer chips*, are in devices great and small that we use every day. Have you ever wondered what is inside those chips? You're about to find out.

Integrated Circuits

Anything that is "integrated" is made of different parts, whether they are people or electronics that are intended to work together as a cohesive unit of some sort. In our case, integrated circuits (ICs) are the equivalent of taking circuits like the ones created on breadboards in the previous projects of this book, shrinking them down to a size that will fit on a small wafer of silicon, and then constructing those miniature components simultaneously. Considering that, a single IC can have numerous capacitors, resistors, diodes, and the like, resulting in devices such as a microcomputer or an AM/FM radio on a single chip.

How is this possible? Here is an extremely simplified explanation: the circuits are built layer on layer on top of a silicon chip. A layer of conductive or doped material is covered by a layer of photosensitive emulsion. A mask with certain areas cut out is then placed over the emulsion and subjected to light, which hardens the exposed areas. The other, unhardened areas of emulsion are then washed away, leaving the base material in the desired spots. This allows the next process to either etch away or deposit more material in only the areas without the emulsion. The remaining emulsion is removed, and then the process begins again, building as many layers with as many different masks as necessary to create the components needed on the silicon chip.

A popular IC is the 555 timer. As its name implies, the 555 timer is used for timing a reaction (power on an output pin) to an input (from an input pin). Other popular ICs are amplifiers, microcontrollers, memory, processors, decade counters, and digital logic chips. Figure 13.1 shows a block diagram of a typical layout for the components on the 555 timer. The triangles represent operational amplifiers (op-amp), each of which has multiple transistors and resistors of its own. The rectangular symbol represents a logic gate called a *flip-flop*, which again will have multiple components in it.

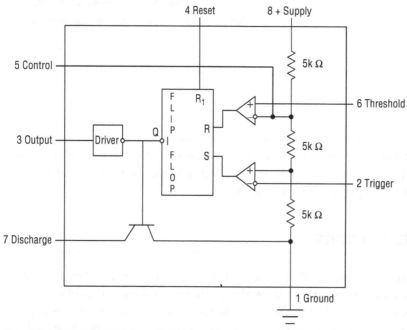

Figure 13.1: Inside a 555 timer IC

The timer has two modes called *astable* and *monostable*. The prefix *a-* means "not," so in astable mode, the output is not stable and is triggered repeatedly until stopped. If the output of the 555 timer in astable mode were graphed, the result would be a square wave, where the voltage is either high or low and switches back and forth for a predetermined amount of time.

The prefix *mono-* means "one," so in monostable mode, a voltage on the output pin is triggered just one time, based on some action on the input pin. It could be used as a delay, perhaps in turning off a light. It could trigger an alarm that will stay on until the circuit is reset, or it could be used to trigger some other timer.

> **NOTE**   Integrated circuit chips are sensitive to static discharge. Less than 100V of static can destroy a chip, and feeling a static shock can be thousands of volts, so take care to eliminate static in the environment and handle the chips carefully. Humidity should be about 50%. Avoid pet hair, carpeting, and fuzzy sweaters that can gather static. Dissipate your static by touching something metal like a doorknob or metal chair leg before you touch an IC. Most are shipped in an antistatic tube or foam and an antistatic bag. After removing the chip from its packaging, place it on an antistatic mat if one is available. Never place ICs on top of an antistatic bag—they should be in the bag, on the breadboard, or on an antistatic mat. When passing an IC to someone else, touch the person first to put you and them at the same electrical potential and then pass the chips, please. Chips are continually improving, so newer chips are less sensitive to electrostatic discharge (ESD) than older chips, but they should still be handled with care.

Try This: Creating an Astable Multivibrator

Again, the best way to learn is by doing, so we'll create a simple 555 timer circuit in astable mode. A multivibrator is essentially an oscillator, which varies from one state to another repeatedly like a grandfather clock's pendulum or a musical metronome. In the case of electricity, the voltage is changing back and forth from a high value to a low value, or in alternating current, from a positive to a negative value as it changes direction.

In an earlier project, a multivibrator was created with two transistors. With the 555 timer, we can achieve the same thing with a couple of resistors, a capacitor, and some jumper wire.

For this project, you need the following materials:

- Breadboard
- Jumper wires, assorted sizes
- 9V battery with snap or other 9VDC power supply
- NE555 timer IC
- 1k-ohm resistor
- 2.2k-ohm resistor
- 220-uF electrolytic capacitor
- LED, any color
- 330-ohm resistor

Optional:

- Assorted capacitors
- Assorted resistors
- Calculator

Determining Circuit Timing

The rate of oscillation is typically an important factor when creating a multivibrator circuit. Luckily, it's not a guessing game and can be easily determined by a formula. The time that the output is high, typically 5V, is called *mark time*, and the time it is low is called *space time*, or *time high* and *time low*.

The frequency of the oscillations can also be calculated. As mentioned in a previous chapter, the frequency is the number of times per second that a complete cycle of a wave is created. A cycle is a complete set of values of the wave from a point, through all its possible values, and back to the same point. See Figure 13.2.

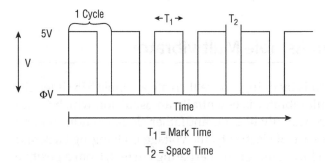

Figure 13.2: Mark time, space time, and cycle

The following are the formulas for mark time (T_1), space time (T_2), and frequency. For mark time and space time, the answer is always expressed in seconds (s). For frequency the unit is hertz (Hz), which means cycles per second. As the resistor numbers get larger and the capacitor numbers get smaller, it's usually easier to work with exponents. See cliffjumperTEK.com/exponents if you need help multiplying exponents, or see cliffjumperTEK.com/electronic_prefixes for help with determining the digital value of a component.

$$T_1 = 0.694 \times \left(R_1^+ R_2 \right) C_1$$

$$T_2 = 0.694 \left(R_2 C_1 \right)$$

With the components outlined in the materials list, using the 2.2k-ohm resistor as R_2, the timing would be as follows:

$$T_1 = 0.694(1000 + 2200)0.00022$$

$$T_1 = 0.5s$$

$$T_2 = 0.694(2200)0.00022$$

$$T_2 = 0.34s$$

As you can see from the calculations, the circuit will output high for 1/2 of a second and output low for 1/3 of a second, which means if it is connected to an LED, the LED will emit light longer than it is off.

Now consider the frequency. The formula for frequency is as follows:

$$F = 1.44 \div \left(R_1 + (2 \times R_2)\right)C_1$$

For our example, frequency is as follows:

$$F = 1.44 \div 1000 + (2 \times 2200)0.00022$$

$$F = 1.21 \ Hz$$

A frequency of 1.21 Hz means that the cycle will complete 1.21 times a second. Imagine a frequency of 1Hz connected to a circuit that drives an output to a seven-segment LED, or four seven-segment LEDs. With a few other logic chips, this could be a digital clock!

The duration of the cycle, known as the *period*, is how long it takes for one cycle to complete. T = period, and it is expressed in seconds. If the circuit completes 1.21 cycles per second, we know one cycle takes less than a second. The formula for period is as follows:

$$T = \frac{1}{F}$$

Incidentally, the frequency and period are inversely proportional, so when solving for the frequency of a wave, divide 1 by the period of the wave. For this project, the period takes about 8/10 of a second, as shown here:

$$T = \frac{1}{1.21}$$

$$T = 0.83$$

One more value to consider in a circuit such as this is the *duty cycle*. Duty cycle refers to the percentage of time that a circuit output is working (i.e., on duty) as compared to when it is off. This becomes more important when talking about pulse width modulation (see Chapter 14, "Pulse Width Modulation") so we'll take a closer look there.

Examining the IC

Looking at an IC in a schematic can be very different than looking at an IC in real life. When looking at a schematic, the pin numbers of the IC may be relocated from what they are physically to make viewing the electrical connections easier. See Figure 13.3. Notice that the pin numbers don't seem to go in a logical order, but the circuit appears neat and easy to follow.

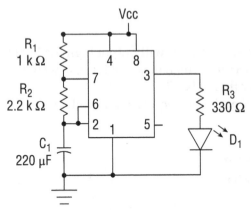

Figure 13.3: A 555 Astable timer circuit schematic

Now look at Figure 13.4. An IC will have some type of indicator on it to denote where pin 1 is. Often it is a divot in one end, but pin 1 could also be noted with a white triangle or a circle. With the divot on the top, the pins will be numbered counterclockwise starting with the top-left pin. When in doubt, it is always best to locate the pinout for the particular chip being used to make sure the assumed pin numbers are correct.

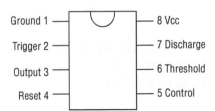

Figure 13.4: A 555 timer pinout

Notice that on the 555 timer chip, pin 1 goes to ground. The ground pin on an IC should always be connected first to avoid damaging the chip. An IC should be placed across the dip in a breadboard so that the left and right sides of the chip are not connected electrically. The connection happens inside the IC.

Building the Circuit

Now it's time to build the circuit.

NOTE   Don't use excessive force to insert an IC into a breadboard. If it doesn't go in fairly easily, you may need to bend the pins slightly toward the center of the chip so they will line up better with the holes they're being inserted into. If an IC must be removed from a breadboard, be sure to lift straight up so as not to bend the delicate pins. There are commercial chip inserter and extractor tools; however, I find the end of the wire stripper, which has a rippled surface, to be a great tool for inserting and removing chips.

Follow these steps to build the circuit. Refer to Figures 13.3, 13.4, 13.5, and 13.6 to aid in placement of the components. The connections referenced here are for the sample board, so adjust the connections to match your board as necessary. If you're using a different 555 timer, remember to check the datasheet and make sure the power is appropriate.

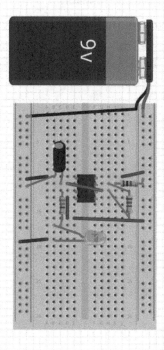

fritzing

Figure 13.5: A 555 timer circuit

1. Insert the 555 timer IC so that pin 1 is in 10e and pin 8 is in 10f, across the dip in the breadboard. Be sure to read the note about inserting the chip into and removing it from the breadboard. Use the positive rail on the side of the breadboard closest to pins 5 through 8 and use the negative rail on the side of the breadboard closest to pins 1 through 4.

2. Jumper from pin 10a to the ground rail, connecting pin 1 to ground.

3. Jumper from 11c to 12g, connecting pin 2 and pin 6.

4. Place the capacitor with the positive lead in 11b and the negative lead in the ground rail.

5. Place a 330-ohm resistor between 12c and 17c.

6. Insert the positive lead of an LED in 17b and the negative lead in the ground rail.

7. Jumper between 13d and 16d, and from 16e to the V_{IN} rail, connecting pin 4 to positive power.

8. Insert one end of the 2.2k-ohm resistor in 12h and the other end in 11h, connecting pins 6 and 7 via the resistor.

9. Insert one end of the 1k-ohm resistor in 11i and the other end in the Vin rail, connecting pin 7 to positive power via the resistor.

10. Jumper between 10j and the Vin rail, connecting pin 8 to positive power.

11. Connect the negative lead of a 9V source into the negative rail and the positive lead into the positive rail. The LED should start blinking.

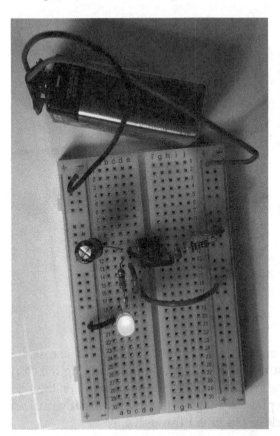

Figure 13.6: A completed astable circuit

Look again at Figure 13.1. The 555 timer has a voltage divider consisting of three 5k-ohm resistors connected in series. The voltage divider provides a consistent reference voltage for the two op-amps, which are also called *comparators*.

Notice that one op-amp is connected to pin 2, which is called the *trigger*. The other op-amp is connected to pin 6, the threshold.

Assume the circuit is at a point where the voltage on pin 2, the trigger, falls below 1/3 of the voltage in. This sets the flip-flop, and the output (pin 3) goes high. Looking at Figure 13.1, notice that the value on the + pin of the comparator connected to pin 2 should be 1/3 of the input voltage because of the voltage divider network, so when the voltage on pin 2 is less than 1/3, the comparator compares the values on its two input pins. And, because the + pin is higher, a signal is sent out of the comparator to the flip-flop, setting it to output high on its pin Q1. This is how an op-amp works as a comparator. This action turns off the transistor connected to pin 7, the discharge pin.

At that point, the capacitor no longer has a direct way to discharge, and it begins charging through R_1 and R_2. When it exceeds 2/3 of the voltage in, the other op-amp reads the input on the threshold pin as higher than the 2/3 of voltage in on the negative op-amp pin and sends an output to the flip-flop that resets the flip-flop and sets the output on pin 3 low. This also turns on the transistor, causing the capacitor to discharge through R_2 and pin 7, the discharge pin. At a point when the voltage across the capacitor again sinks below 1/3 of the input voltage, the process starts all over again.

The capacitor charges through R_1 and R_2, but it only discharges through R_2 to pin 7, which is why the calculations for mark time and space time are different.

Pin 5, the control pin, is used in a situation where the values for the control side of the op-amp need to be different than the normal 1/3 and 2/3.

Pin 4 can be used to reset the timer.

The following is a recap of the pins of a 555 timer:

1. Ground
2. Trigger
3. Output
4. Reset
5. Control
6. Threshold
7. Discharge
8. Power supply

Variations of the 555 timer are a 556 timer that has two timer circuits on one IC, and 558/559 timers that have four timers on one IC.

By connecting a speaker and varying the frequency, an output sound can be emitted. Using two timers together at different frequencies can create a warbling sound effect like a police siren. For more fun 555 timer circuits, visit `cliffjumperTEK.com/555_timer`.

Operational Amplifiers

Operational amplifiers (op-amps) can do more than simply compare two inputs, as discussed in the astable multivibrator circuit. As their name implies, they can also be used to amplify a weak signal and make it stronger.

An op-amp is identified in schematics as a triangle. Two pins of the op-amp are input. One is called *inverting* and the other *noninverting*. There is one output pin, and like any IC, it needs power to run the chip and a ground pin. An op-amp will amplify the difference between the inverting and noninverting input pins. When used as an amplifier, quite often one input or the other will be attached to ground (0 volts) instead of having an applied voltage. Because no electronic device is completely perfect, an op-amp has two offset pins that are used to fine-tune the input pins, so without an applied voltage, they will truly be at zero volts.

Check out Figure 13.7. The center image shows a typical pinout for an op-amp. The top image shows an op-amp configured as an inverting amplifier. Power is applied to the inverting input pin, which is usually pin 2. The noninverting pin, pin 3, is connected to ground. R_I stands for input resistor, and R_F stands for feedback resistor. The gain on the op-amp can easily be calculated. There is no unit for gain because it is a factor that compares input to output. Its symbol is the letter A, for amplification. The formula for gain in an inverting op-amp is $A = -\dfrac{R_F}{R_I}$. The minus sign is in the formula because the output is inverted from the input.

An op-amp can also be wired noninverting, so the output has the same polarity as the input. In Figure 13.7, the bottom circuit is noninverting. The formula for amplification in a noninverting op-amp circuit is $A = 1 + \dfrac{R_F}{R_I}$.

In either case, whether inverting or noninverting, the output voltage will be equal to the input voltage multiplied by the gain.

Digital Logic

When examining a computer, calculator, or other digital device's hardware, it becomes apparent that the only thing the computer "knows" is that either a particular wire, called a *trace*, has power on it or it does not. Digital logic is the process that compares those on and off signals and can make decisions based on what is there.

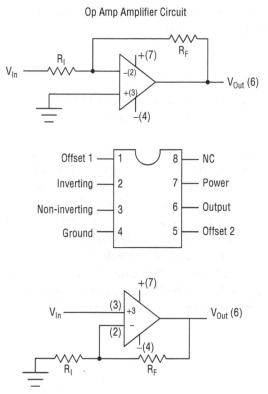

Op Amp Amplifier Circuit

Figure 13.7: Inverting and noninverting op-amp

Logic Chip Construction

Digital logic depends on voltage, and as mentioned in Chapter 4, "Introduction to the Arduino Uno," logic high is roughly 5V and logic low is 0V. However, what if the voltage is not exactly that? It depends on the construction of the chip.

The two basic types of ICs are transistor-transistor logic (TTL) or complementary metal-oxide semiconductor (CMOS). TTLs operate on roughly a 5V input, with a small acceptable variance. A voltage reading between 2 and 5 is considered logic high, voltage between 0 and 0.8 is considered logic low, and 0.9 to 1.9 is considered the invalid range. The output range is only slightly different, with 0.6V to 2.6V being the invalid range. TTL chips are fast and come in low- and high-power varieties.

CMOS chips have better noise resistance than TTL chips; however, they are more sensitive to static discharge and must be handled carefully. CMOS chips can come in high-speed varieties. The big advantage of CMOS chips is that they can operate on voltages as high as 15V, depending on the chip being used. If a CMOS is operating on 5V, for inputs logic high is 3.5V to 5V, and logic low is 0V to 1.5V. To know the exact logic range for a chip, refer to the chip's datasheet.

TTL chips are based on bipolar junction transistor (BJT) technology, while CMOS chips are based on field effect transistors (FETs).

The Binary Number System

Binary numbers were discussed in Chapter 4, but it is important to iterate that that logic high will equate to on or binary 1, while logic low equates to off or binary 0.

Binary numbers can be converted to decimal numbers using a computer's programming calculator; however, understanding the background is helpful. Consider that a decimal number, which is the numbering system we use and which our money is based on, has 10 characters (0–9) available. The decimal number system increments by multiples of 10. For example, going from right to left, the values of placeholders in a decimal number system would be 1, 10, 100, and 1,000. Each successive placeholder is 10 times the one before.

Binary numbers increment based on multiples of 2, because only 0 and 1 (off and on) are available. Going from right to left in the binary numbering system, the values of placeholders are 1, 2, 4, 8, and so on. See Figure 13.8.

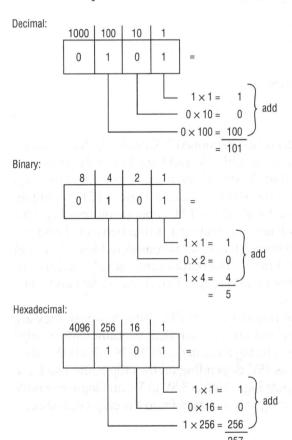

Figure 13.8: Decimal, binary, and hexadecimal

Binary numbers in a computer system are used to represent everything from numbers and letters to line feed characters and colors. American Standard Code for Information Interchange (ASCII) is a standardized code that programmers use so that devices are able to communicate effectively. For example, the number 9 is represented by 00001001, while the letter *a* is expressed as 0101001, and 00010001 represents the letter *A*. A colon (:) is represented by 00001010, and so on. By using ASCII or the more modern Unicode, digital systems are able to translate data between the binary that they understand and human-readable characters.

Sometimes, when the numbers are very large, computing systems represent them to us as hexadecimal (hex) values. The hexadecimal number system is base 16, represented by the characters 0 through 9, plus A, B, C, D, E, and F. A is equal to 10, B is equal to 11, and so on. As with binary and decimal, each subsequent placeholder is 16 times the one before. 0x is sometimes used before a hexadecimal number to indicate that it is indeed hex. 0xA5 would be $(10 \times 16) + 5$, or the decimal equivalent 165. See Figure 13.8 for a comparison of numbering systems. For more work with binary and hexadecimal, visit `cliffjumperTEK.com/binary_is_fun` or `cliffjumperTEK.com/hex_is_fun`.

Logic Gates

Some smart engineers have created wonderful ICs for us called *logic chips*. Logic chips read the input on one or more pins and respond with an output of logic high or low, depending on what they read there. These chips are called *logic gates*, because they provide a way to use Boolean logic (is it true or false) in electrical circuits.

Take, for example, a 2 input AND logic gate. Assume the two inputs are represented by A and B, and the output is C. If A is logic high, and B is logic low, then C (output) would be logic low, because with an AND chip both inputs A and B must be high for the output to equate to true. If A is logic low, and B is logic low, then C would be logic low. If A is logic low and B is logic high, then C would be logic low. If A is logic high, and B is logic high, then C would be logic high. With an AND chip, all inputs must be high to get a high output.

Luckily, there is a simpler way to explain all the inputs and outputs, and that is with a truth table, and the AND gate is certainly not the only one. Other logic chips are OR, NOT, Hex Buffer, NAND (pronounced N-AND), NOR (pronounced N-OR), XOR, and XNOR. The N in a logic gate name means not, and the X stands for exclusive. The NOT gate is also called an *inverter* because the output is inverted from the input. NOT and Hex Buffer are single input gates. Figure 13.9 shows the truth tables for these logic gates.

The truth tables for the two input logic gates in Figure 13.9 have been combined into one to save space. To read the table, the two inputs are on the left, and all outputs are on the right. If, for example, the circuit required an OR gate and the two inputs were 0 and 1, following across to the OR column, the output would be 1.

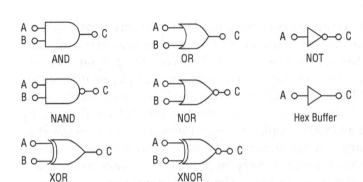

Combined Truth Table

Input		Outputs					
A	B	AND	OR	NAND	NOR	XOR	XNOR
0	0	0	0	1	1	0	1
0	1	0	1	1	0	1	0
1	0	0	1	1	0	1	0
1	1	1	1	0	0	0	1

Single Input Gates

Input	Output	
A	Not	Buffer
0	1	0
1	0	1

AND

3 Input			Out
A	B	C	Q
0	0	0	0
0	0	1	0
0	1	0	0
0	1	1	0
1	0	0	0
1	0	1	0
1	1	1	1

Figure 13.9: Logic gates and truth tables

Here is brief explanation of each type of logic gate:

- **AND:** All inputs must be high for the output to be high.
- **OR:** If any input is high, the output is high.
- **NAND:** Outputs are the opposite of what they would be with an AND chip.
- **NOR:** Outputs are the opposite of what they would be with an OR chip.
- **XOR:** Either A or B high, but not both equates to output high.
- **XNOR:** Opposite of XOR.

- NOT: It inverts the input. Low = High, High = Low.
- Hex Buffer: Does not invert the input. Low = Low, High = High.

Notice in the combination truth table in Figure 13.9, only two inputs are used, but most gates can have more than two inputs, as with the separate three-input AND gate truth table shown. If a gate has more inputs than will be used, the unused input should be either connected to another input or connected so that it is always "pulled up" to logic high or "pulled down" to logic low, depending on the desired outcome. Logic gates can also be combined into wonderfully complex designs.

In addition to having more than two inputs, a single IC can have more than one logic gate on it, and in fact many have four or more. See Figure 13.10.

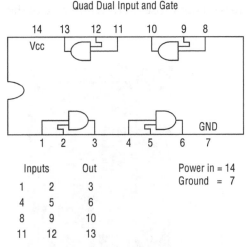

Figure 13.10: A logic gate pinout

Try This: Exploring AND and OR Gates

There are a few tips and tricks to using logic chips. They can, on occasion, be finicky creatures; however, you will know just what to do to make them work well. Practice connecting them now, and you'll be more confident and efficient when you need to do it on a bigger project.

For these projects, the following materials are needed:

- SN74HC08N four-gate, two-input AND chip
- SN74HC32N four-gate, two-input OR chip
- Full-size breadboard or more than one smaller breadboard
- (2) switches

- 1200-ohm resistor
- LED, any color
- (2) 330-ohm resistors
- 4.5V–6V power source with leads
- Assorted jumper wires

The AND Gate

First, build the circuit using the AND gate. Follow the instructions in this section and refer to Figure 13.11 to help you with your circuit. These instructions reference points on the demo board and may need to be adjusted to match your board. The circuit shown is somewhat spread out to make it visually easier to follow. The terminal row numbers referenced are those on the right side of the board.

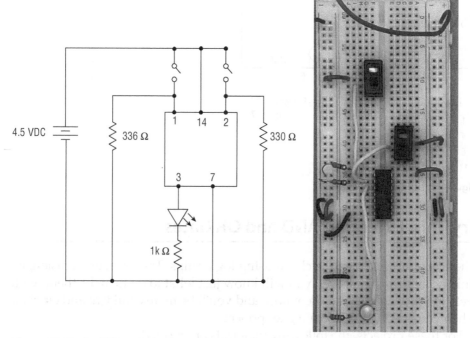

Figure 13.11: An AND gate pinout and circuit

1. Jumper across the board connecting the power rails on each side to the other (i.e., red to red and black to black), so ground and positive power can both be accessed on either side of the board.

2. If the colored strip for the power rails is split in the middle, jumper across the split so the top and bottom of the breadboard power rails are connected. Remember to do this on both sides of the breadboard.

3. Place one switch so the common connector is in pin 10f and the other connector is in pin 11f.

4. Place the second switch so the common connector is in 20c and the other pin is in 21c.

5. Place a red jumper between the closest power rail and 10j (connecting power to the common pin of the first switch).

6. Place a red jumper between the closest power rail and 20a (connecting power to the common pin of the second switch).

7. Carefully remove the AND chip from its antistatic packaging and orient it so that the divot is facing toward the switches. Pin 1 will be on the top left and pin 14 on the top right. Insert the IC so that pin 14 is in 25e and pin 1 is in 25f across the dip in the breadboard.

8. Jumper from 31j to the ground rail (connecting the GND pin, pin 7, to ground).

9. Jumper from 25a to the closest power rail (connecting pin 14 to power).

10. Place a 330-ohm resistor so that one end is in 25j and the other end is in the ground rail.

11. Place a 330-ohm resistor so that one end is in 26j and the other end is in the ground rail. (This step and step 10 connected pins 1 and 2 to ground.)

12. Connect a jumper between 11i and 25i (connecting the first switch to pin 1).

13. Connect a jumper between 21e and 26g (connecting the second switch to pin 2).

14. Connect a jumper between 27g and 45g.

15. Insert the long lead of an LED in 45h and the short lead in 46h.

16. Insert a 1200-ohm resistor between 46j and the ground rail (connecting the negative lead of the LED to the ground rail).

17. Insert the positive and negative leads into the appropriate power rails.

Turn on both switches. The LED should light. If it doesn't, check all of your connections. If the AND chip is getting hot, disconnect the power immediately and check to ensure that it is not in backward and that the power source is appropriate for the AND chip you're using.

With the LED lit, if either switch is turned off, the LED should turn off.

The 1,200-ohm resistor is there to limit current through the LED.

Perhaps you're wondering why there are resistors connecting pin 1 and pin 2 to ground. Without them, the circuit may or may not work; typically it will be inconsistent. The resistors move any extra circuit noise, which is essentially static electricity, to ground. Without these resistors the circuit might work, but you may notice that if you move your finger near the chip, the static on your hand would be enough to make the light turn off or on; in fact, any stray voltage may be enough to activate it. Adding the resistors requires a more determined signal to affect the gate.

Pull out the two resistors connecting pin 1 and pin 2 to ground. Turn the switches off and on and observe what effect removing the resistors has on the circuit.

Reinsert the resistors. Turn off the switches and unplug the power source from the power rail. Leave the circuit intact for the next project.

The OR Gate

Refer to the OR gate chip pinout in Figure 13.12. Notice that the power pin is pin 14 and the GND pin is pin 7, the same as the AND chip. Pins 1 and 2 are inputs, and pin 3 is an output. This is an OR gate, so its truth table is different from the AND gate.

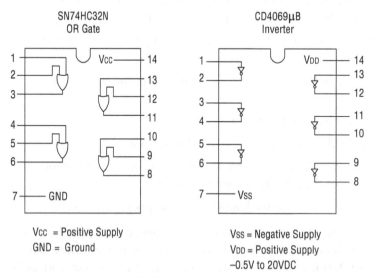

Figure 13.12: OR gate and NOT (inverter) pinouts

Follow the instructions in this section and refer to Figure 13.12 to help you with your circuit. These instructions reference points on the demo board and may need to be adjusted to match your board. The terminal row numbers referenced are those on the right side of the board.

1. Ensure that the power source is disconnected from the breadboard to avoid damaging the chips.

2. Gently pull straight up on the AND chip, removing it from the breadboard, and place it immediately in its antistatic packaging.

3. Orient the OR chip so that the divot is on the top facing the switches, with pin 1 on the top left and pin 14 on the top right.

4. Insert the OR chip where the AND chip was, with pin 1 in 25f and pin 14 in 25e.

That's it! Now that the OR chip is being used, the LED should light if either one of the switches is turned on. It will also light if both switches are turned on. Experiment with this new configuration.

Logic Probes and Oscilloscopes

In the circuits shown previously, an LED was used to indicate power coming from an output pin, but what if there was no indicator light? A *logic probe* can be used to determine logic high or logic low on an IC pin. The probe has a pointed metal end and alligator clips. The alligator clips are attached to the circuit for power, and then the probe is touched to an IC lead. One or more lights are used to indicate the voltage high or low condition. Some have a switch to change between TTL and CMOS IC construction.

A similar but opposite device is a *logic pulser*. See Figure 13.13. A logic pulser is used to inject a signal into a logic circuit for testing purposes. The advantage of a pulser is that components don't have to be removed from the circuit to be tested. Logic pulsers and probes look similar to each other, and inexpensive kits can be purchased to solder your own logic probe or pulser.

Oscilloscopes are another handy tool to have when working with circuits. Oscilloscopes will show voltage measurements like a multimeter does but provide much more information. A typical oscilloscope will show voltage, duty cycle, frequency, and even the waveform created by a circuit.

Figure 13.14 shows an inexpensive oscilloscope measuring the voltage changes over time, across the capacitor of a 555 timer circuit, resulting in a sawtooth pattern. Notice the readings available on the top left of the oscilloscope. It provides frequency, period, pulse width, and duty cycle, along with the voltage measurements on the right.

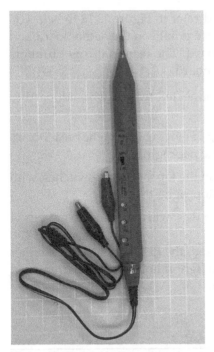

Figure 13.13: Logic pulser

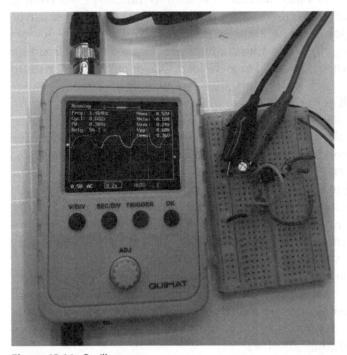

Figure 13.14: Oscilloscope

An oscilloscope will have at least one input, which is usually connected with a BNC connector. (BNC stands for British Naval Connector, although some people call it a Bayonette Connector.) To attach the lead using the BNC connector, push down while turning the connector. The input for this oscilloscope is at the top left. The opposite end of the connector has two alligator clips. Others may have a different mechanism for connecting to a circuit. Connect the black lead to the more negative side of what is being measured first; then connect the red lead on the positive side.

A dual trace oscilloscope has two inputs and is handy for comparing wave signals. Each signal will be displayed as a different color on the screen. Some oscilloscopes have four or more inputs.

Often an oscilloscope needs to be calibrated before it is used. The exact process varies by oscilloscope and should be detailed in the user's manual. Calibration adjusts for the resistance in the leads, so often a lead will need to be connected to a test point on the oscilloscope and a certain waveform achieved each time the oscilloscope is turned on for use.

Like many graphs, the output of an oscilloscope is a grid of squares indicating vertical and horizontal values. Going from left to right the horizontal grid measures time, and from bottom to top the vertical grid measures voltage. Each grid square is called a *division*. Just below the output screen is a row of buttons. The black button on the left says V/DIV, which stands for volts per division. Pressing this button and then turning the adjustment knob allows the user to change how many volts are represented by each vertical square. In this particular example, each division is 0.5V. If an output should be a wave form but appears as a flat line, the volts per division may need to be adjusted.

The next button to the right says SEC/DIV, which stands for seconds per division. Just as it sounds, pressing this button and turning the adjustment knob will allow the user to change what part of a second is represented by each horizontal square. In this example, each division represents 0.2 seconds, so five divisions are 1 second. Examining the waveform, we can see that it takes just over three divisions for a wave to complete (3 divisions × 0.2 seconds = 0.6 seconds). Looking above at the cycle reading, we can see that the period of this wave is indeed measured at 0.602 seconds. Many oscilloscopes have an auto-adjustment feature that will determine automatically the best volts and time per division. Think of using the adjustments as zooming in or out in voltage and time.

Oscilloscopes also have a trigger feature. The trigger feature tells the oscilloscope at what point to start drawing the waveform on the screen. Perhaps it is set to 0 volts. The oscilloscope will start drawing the wave when it reaches 0 volts. If not for the trigger, the oscilloscope output would be very disorganized and appear as a blur because the waveforms wouldn't line up properly.

Another common oscilloscope feature is called *hold*, which takes a snapshot of the screen at a given point and keeps it on the screen.

Some oscilloscopes have the capability of being connected to a computer via a USB port so that the information can be stored in computer memory. It's especially important to understand the proper use of such oscilloscopes to avoid damaging the computer they are connected to.

Oscilloscopes range in price from around $50 for an inexpensive handheld one, like the one in Figure 13.14, to tens of thousands of dollars for sophisticated oscilloscopes.

More Please

In This Part

Pulse Width Modulation

"I'm 27 and I have no idea what I am doing half the time. I am just trying the best I can and I think that speaks to a lot of other women out there."

—Whitney Wolfe Herd

I just love that quote. Like Whitney, we may not always know what we're doing, but if we keep doing something in the right direction even part of the time, things will average out great, especially over the long term.

Pulse width modulation (PWM) is like that. PWM has to do with duty cycles and doing something only part of the time, which averages out to a certain voltage overall, and can do things such as create incredible light displays, control the speed of your hovercraft, and vary the power to dual motors more, or less, to turn your RC car from left to right.

Pulse Width Modulation Explained

Many of the analog devices in our lives are controlled by digital circuits. We see certain wavelengths (analog signals) of the electromagnetic spectrum as color, which can be generated by our digital TVs. The sound output of our phones is an analog signal controlled by a digital device. The speed of forward motion of tires on a road is analog, controlled in part by digital circuits in a car, or in the case of a Tesla, completely by electrical circuits. Seemingly everywhere in our world, analog signals are being converted to digital and digital to analog. One of the ways this is done is by using PWM. PWM has everything to do with duty cycles, which were introduced in Chapter 13, "Integrated Circuits and Digital Logic," along with the 555 timer and the oscilloscope.

The duty cycle refers to the comparative length of time that a digital circuit is on as opposed to off. When we use a 555 timer, the mark time (on) and space time (off) can be easily calculated and changed based on the components we attach to the circuit.

Figure 14.1 illustrates what the output of an oscilloscope would be in the following situations. Assume that with our 555 timer we created a circuit where mark time and space time are equal. That would yield a duty cycle of 50%, because the output voltage would be high half of the time and off half of the time. If the power supplied on that circuit is 9VDC, with a duty cycle of 50%, then the *average* of the voltage over time would be 4.5VDC. (This is not to be confused with the average of an AC sine wave. The AC sine wave is an analog signal where the average is calculated differently.) PWM uses a digital signal, where the signal is either completely on or completely off (in a perfect situation) to mimic an analog signal. As a refresher, an analog signal is more like a wave, where there exists a complete range of values. Digital signals have but two values—on and off.

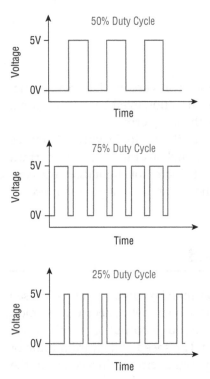

Figure 14.1: Duty cycles

If we adjust the components of the 555 timer circuit such that the duty cycle is 25%, then the average voltage over time would be $0.25 \times 9V$, equaling 2.25V. If we created a 75% duty cycle, the average voltage output would be 6.75VDC. What if we wanted an average of 90% power or 10% power? Simply adjust the components of the circuit to convert the digital signal to the appropriate duty cycle. PWM can be used to vary the average voltage and mimic an analog output while it is controlled by a digital device, such as a computer program or electronic circuit. Replace one of the fixed resistors in the 555 timer circuit with a variable resistor attached to a knob, and the duty cycle can be changed in an instant by simply turning the dial.

If a digital signal with a lamp as its output has a mark time and space time of 1 second each, it has a 50% duty cycle. But because each on/off cycle is so long, we would be able to observe the circuit turning off and on. If PWM were being used for a project like dimming a light, the frequency would need to be high enough so that it no longer appeared to be going on and off as a digital signal would but fast enough so that we could not see it turning off and on. Hence, PWM requires attention to duty cycle *and* frequency.

An important consideration with PWM is the choice of components. If a component had a maximum peak voltage of 5VDC, it would not be appropriate for a circuit where the maximum peak voltage will be 9VDC. The component would likely overheat and be damaged or worse. This is another time when reviewing the datasheet of the component is important, regardless of what the component on the output of the circuit is. Many datasheets will list the typical voltage that a component will operate at safely, as well as a maximum voltage and what duty cycle or percent of time the component can be at that high voltage. Most will also have a sustained maximum voltage, which will be less than the peak. See the datasheet snippet in Figure 14.2.

Now that you know a bit more about PWM, let's build some circuits that use PWM. First, we'll dim some lights; then, we'll control a motor or two. The final project uses an Arduino board to control PWM and give just a taste of its possibilities.

Try This: Using a PWM LED Dimmer

This project employs the familiar 555 timer circuit to teach the basics of PWM. The LEDs could easily be replaced with an appropriate fan, relay, or other device, with an appropriate circuit on the output side. The key in choosing a component is to ensure that it can handle PWM and the maximum voltage.

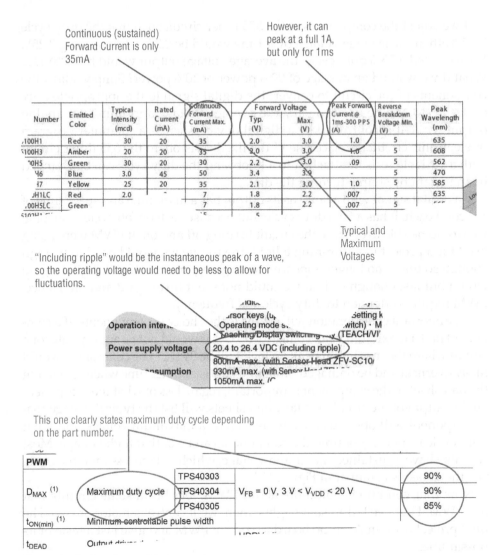

Continuous (sustained) Forward Current is only 35mA

However, it can peak at a full 1A, but only for 1ms

Number	Emitted Color	Typical Intensity (mcd)	Rated Current (mA)	Continuous Forward Current Max. (mA)	Forward Voltage		Peak Forward Current @ 1ms-300 PPS (A)	Reverse Breakdown Voltage Min. (V)	Peak Wavelength (nm)
					Typ. (V)	Max. (V)			
.100H1	Red	30	20	35	2.0	3.0	1.0	5	635
5100H3	Amber	20	20	35	2.0	3.0	1.0	5	608
0H5	Green	30	20	30	2.2	3.0	.09	5	562
46	Blue	3.0	45	50	3.4	3.9	-	5	470
47	Yellow	25	20	35	2.1	3.0	1.0	5	585
JH1LC	Red	2.0	~	7	1.8	2.2	.007	5	635
.00H5LC	Green			7	1.8	2.2	.007	5	

Typical and Maximum Voltages

"Including ripple" would be the instantaneous peak of a wave, so the operating voltage would need to be less to allow for fluctuations.

Operation inter...	rsor keys (u, Operating mode s. Teaching/Display switching ...	Setting k .witch) · M y (TEACH/VI'
Power supply voltage	20.4 to 26.4 VDC (including ripple)	
...nsumption	800mA max. (with Sensor Head ZFV-SC10/ 930mA max. (with Sensor H...) 1050mA max. /(...	

This one clearly states maximum duty cycle depending on the part number.

PWM				
D_MAX (1)	Maximum duty cycle	TPS40303	V_FB = 0 V, 3 V < V_VDD < 20 V	90%
		TPS40304		90%
		TPS40305		85%
t_ON(min) (1)	Minimum controllable pulse width			
t_DEAD	Output driv...			

Figure 14.2: Datasheet with duty cycle

PWM vs. Potentiometer

This 555 timer circuit will use a potentiometer to facilitate changing the duty cycle. You may wonder, "Why not just use a potentiometer to dim the lights?" The answer is simple. PWM provides the full power in brief spurts, where a potentiometer, when it decreases the resistance, will increase the circuit current. For some devices, this is critical because the component could be damaged. Too much current on a conductor can also cause overheating of the conductor and perhaps melt its insulating coating. In PWM, it's as if the circuit is simply being turned on and off again very fast. The values on the circuit aren't actually changed, just the duration.

Building the Circuit

For the PWM LED dimmer circuit, the following components are needed:

- Full-size breadboard
- (2) switches
- 555 timer
- 1k-ohm resistor
- 100k-ohm potentiometer
- (2) 0.01uF capacitors, mylar or ceramic disc
- (2) 1N4007 or similar diodes
- (2) 9VDC source with leads
- (2) LEDs, preferably blue
- 220-ohm resistor
- 330-ohm resistor
- 2N5551 NPN transistor or similar
- Oscilloscope (recommended but not required)

Look first at the right side of the circuit in Figure 14.3, the output side. Two blue LEDs have been chosen because they each can transduce 3.0V to 3.4V of electricity to light, and using two of them with the current-limiting 220-ohm resistor means that they will not be overpowered by the 9V source.

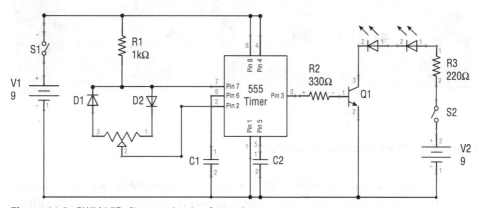

Figure 14.3: PWM LED dimmer circuit schematic

The transistor used is a BJT, and it's needed to interface between the input and output sides of the circuit because the output from pin 3 of the 555 timer provides enough voltage to "turn on" the transistor but not enough to run the load (the two LEDs and resistor), which is the reason for the transistor and 9V second voltage source. This particular transistor can support 600mA on the collector, collector-emitter voltage of 160V, collector-base voltage of 180V, and emitter-base voltage of 6V. It can dissipate 625mW of heat, and its maximum frequency is 100 MHz, and I happened to have one on hand. It's more than this particular circuit needs, of course, so you could use a different transistor as long as you check the datasheet to verify that it will be powerful enough for the circuit. For running something larger, you may want to use a transistor with a heat sink, and of course, you would check the datasheet of any component that you're using in your circuit. As a refresher, a small voltage on the base of the transistor effectively "turns on" the transistor and allows a larger current to flow between the collector and emitter of the transistor.

Although this circuit uses a 9V source on both the input and the output sides, the output side could be virtually any voltage and circuit, as long as you ensure that the components are capable of using that voltage.

Let's get the circuit up and running, and then I'll explain how it works. See Figure 14.4 and Figure 14.5 to help you build the circuit. The contact points referenced are the demo board. You may need to adjust the positions for your breadboard.

> **NOTE**   The demo breadboard has split power rails in the middle so it can have two input voltages. If the breadboard you're using doesn't have this capability, you'll need to use two breadboards.

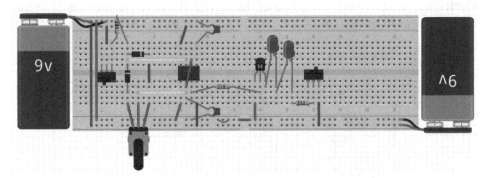

Figure 14.4: PWM LED dimmer circuit pinout

1. On the left end of the breadboard, jumper from the rails on one side to the other so power and circuit ground can be accessed from both sides.

2. Place a switch so that the middle pin is in 6d and one end pin is in 7d.

3. Insert the 1k-ohm resistor with one lead in 7a and the other lead in 10a.

4. Jumper between 6a and the positive rail.

5. Place the anode of one 1N4007 in 20c and the cathode in 14c.

6. Place the cathode of the other 1N4007 in 10e and the anode in 10h, crossing the breadboard dip.

7. Insert the 555 timer so that pin 1 is in 20f and pin 8 is in 20e, across the dip.

8. Jumper between 10d and 21d (between the diodes to pin 7).

9. Jumper between 14e and 14f, crossing the breadboard dip.

10. Insert the leads of the 100k-ohm potentiometer so that the middle lead is in 12j, one side lead in 10j, and the other side lead in 14j.

11. Jumper between 12h and 21h, connecting the center lead of the potentiometer to pin 2 of the 555 timer.

12. Jumper between 14e and 14f, across the breadboard dip and connecting the negative end of one diode to the outer pin of the potentiometer. (Note: The other diode should be connected to the same terminal row as the opposite end of the potentiometer.)

13. Jumper between 20j and the circuit ground (pin 1 to ground).

14. Jumper between 21g and 22d, connecting pin 2 and pin 6.

15. Place one lead of a 0.01uF capacitor in 21j and the other lead in the circuit ground rail, connecting pin 2 to ground through the capacitor.

16. Insert one lead of a 330-ohm capacitor into 22h and the other lead into 35h.

17. Jumper between 23j and the positive rail, connecting pin 4 to positive power.

18. Jumper between 20a and the positive power rail, connecting pin 8 to positive power.

19. Place one lead of the other 0.01uF capacitor in 23a and the other lead in the circuit ground rail.
That completes the left side of the circuit. Now on to the right, the output side.

20. If the ground rail of the breadboard is split, jumper across the split to connect the left ground rail to the right. Please see Figure 14.5.

NOTE   Do *not* connect the two sides of the power rail. While it's true that in this particular circuit both sides can use the same voltage, that will not always be the case. It's better to get in the habit of separating the input voltages now.

21. Insert the transistor so that the emitter is in 34f, the base is in 35f, and the collector is in 36f, causing the 330-ohm resistor and the transistor base to be in the same terminal row.

Figure 14.5: PWM LED circuit complete

22. Jumper between 34j and ground, connecting the transistor's emitter to circuit ground.

23. Place the negative lead of one LED in 36g and the positive lead in 38g.

24. Place the negative lead of the other LED in 38h and the positive lead in 41h.

25. Place one lead of the 220-ohm resistor in 41j and the other lead in 45j.

26. Place switch 2 so that the center lead is in 45g and one outside lead is in 46g.

27. Jumper between 46j and the positive power rail.

28. Insert the leads of one 9V battery into the appropriate rails on the left end of the breadboard and the leads of the other 9V battery into the appropriate rails on the right side of the breadboard. (Refer to Figure 14.5.)

29. (This step is optional but highly recommended.) Strip all the insulation off about 3/4 inch of 22-gauge jumper wire. Bend and insert both ends of the jumper wire into the ground rail, as shown in Figure 14.5. Ensure that there is a loop of wire sticking up enough to connect to it with an alligator clip. This loop of wire is a connection point to ground for your oscilloscope, which you're going to want to use to examine your circuit.

Ta-da! Another circuit completed. Now for the fun part.

Turn both switches on and slowly turn the potentiometer first one way and then the other. The LEDs should light and get brighter and dimmer depending on which way the potentiometer is being turned, but what is happening electrically?

When S_1 is closed, the capacitor, C_1, begins charging through R_1, D_1, and the left side of the potentiometer. C_2 is on the circuit only to smooth out any voltage fluctuations.

As is normal with a 555 timer and explained in Chapter 13, the circuit will flip and C_1 will discharge through D_2 and pin 7. Time on is determined by C_1 and the values of R_1 and the left side of the potentiometer. Time off is determined by C_1 and the value of the right side of the potentiometer. By making the value of R_1 relatively low as compared to the value of the potentiometer, its effect is negligible, and essentially the duration of time on and time off is determined by the position of the potentiometer.

During this charging and discharging, pin 3 will be switching back and forth between output high and output low. When pin 3 is logic high, the voltage is sufficient to allow current to flow between Q_1's collector and emitter, lighting the LEDs. When pin 3 is logic low, the LEDs do not emit light because there is no current flowing in that circuit. The frequency is fast enough that human eyes don't see the on and off of the light but instead see the reduced time on or off as a more or less dim light.

Observing the Changing Duty Cycle

To connect the oscilloscope, see Figure 14.6. Follow the manufacturer's directions to calibrate the oscilloscope if necessary. Turn off the switches on the circuit. Connect the red (positive) lead of the oscilloscope to the lead of the resistor connected to pin 3. Connect the black (negative) lead of the oscilloscope to the loop that was made for this purpose in step 29. Alternately, it needs to be connected to circuit ground in some fashion.

Turn on the circuit and the oscilloscope. Adjust the volts per division and seconds per division until a square wave is shown. Refer to the documentation with your particular oscilloscope for directions on how to adjust settings and use the trigger. This is where an auto setting feature is helpful.

Once the square wave shows on the oscilloscope's screen, turn the potentiometer from left to right and back again, and observe the change in the LED's brightness and the duty cycle (time on) as you do so. If you don't have an oscilloscope handy, you can see the circuit in action on cliffjumperTEK.com/PWM.

It's recommended that this project be kept together on the breadboard to facilitate setting up the next one.

Try This: Using a PWM Motor Control

A PWM circuit can be used to control motor speed in an analog manner. A pair of them can be used to turn a vehicle left or right, depending on which motor is turning a wheel (or fan) faster. A PWM circuit can also control the speed forward, whether it's an RC car motor or the fan pushing a drone. Motors, fans, relays, and other devices that have an inductor (discussed in Chapters 8, "Diodes: The One-Way Street Sign," and 11, "The Magic of Magnetism") need special consideration when creating circuits.

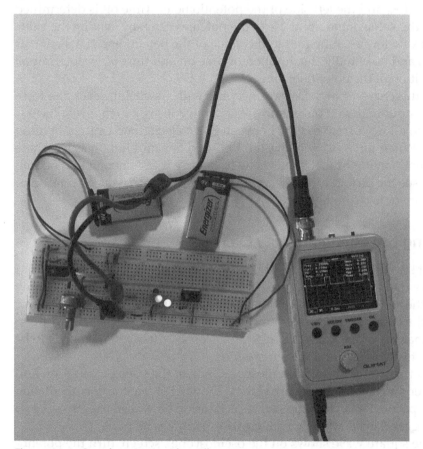

Figure 14.6: Complete circuit with oscilloscope

Inductors store energy in a magnetic field. When current to an inductor is stopped (a switch is turned off), the inductor's magnetic field will collapse and rapidly return the electricity that is held in it to the circuit, causing a voltage that is in reverse to the initial voltage. It manifests as a large voltage spike, which can be sufficient to overcome the resistance of the air between connectors in a switch, causing arcing and damaging the switch, and it can feed back to the source it came from, damaging any components in its path. When a transistor is being used as a switch, as in our PWM circuit, the transistor could be destroyed.

To avoid damaging circuit components, a flyback diode is used to mitigate the spike and protect the components. See Figure 14.7. Conventional flow theory is often used to explain flyback diode functioning, so we'll do the same here. When there is no flyback diode, current flows through the motor and the switch (in our case, a transistor) to ground. When the switch is turned off, the inductor of the motor resists the change in current, meaning that the current must flow out of the inductor as the magnetic field collapses. The problem is that without a complete circuit, the charge has no place to go and there ends up being a positive charge on what was the negative side of the motor. If this charge is sufficient, it will arc across a closed switch, or in our case, damage the transistor. It can also feed back through the positive voltage source as it's trying to find a way to ground.

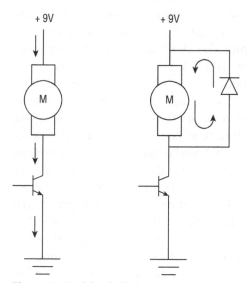

Figure 14.7: Flyback diode

The solution is to add a flyback diode, also known as a suppressor diode or snubbing diode. The diode is in parallel to the motor and reverse biased, meaning that the cathode of the diode is toward the positive side of the circuit. When the switch is closed and current is flowing through the motor to ground, it's as if the diode part of the circuit wasn't there. Remember, the main purpose of a diode is to allow electricity to flow in just one direction, and with the diode reverse biased, it blocks the flow of electrons through that part of the circuit.

With the flyback diode in place, when the switch is opened rather than a positive charge accumulating at the normally negative end of the motor, the charge flows through the diode back to the opposite terminal of the motor, making the diode forward biased for that circuit. The power is safely dissipated without damaging the rest of the circuit.

Other variations exist on this theme, but the premise is the same. The diode allows for safe dissipation of the charge. A circuit may have another diode in parallel with the transistor to further protect it, as well as a capacitor going to ground between the positive source and the motor, so that feedback from the motor will go to ground rather than go on to the source.

There are many different types of diodes, and some are more suitable for specific purposes than others. (For a refresher on the types of diodes, refer to Chapter 8, "Diodes: The One-Way Street Sign.") When selecting a flyback diode, be sure to choose one that is a rectifier diode with a sufficient power rating. Under normal operation, the diode is blocking current from flowing through it, so the peak inverse voltage rating (PIV) needs to be a bit higher than the voltage source. The current of the collapsing magnetic field should start close to the current that created it and eventually trail off to nothing (remember the inductor's transient response from Chapter 11, "The Magic of Magnetism"). So, if the normal current through the motor were 1A, you would want the diode to be able to handle a minimum of 1A. But this is a case where exceeding what is expected is inexpensive and desirable. The voltage rating of the diode needs to be higher than the source voltage due to the spike that will occur. So again, far exceeding what is expected to be on the circuit is inexpensive and desirable.

In the case of a bidirectional motor—one that can be made to reverse direction, depending on which side of the motor is receiving the positive and negative power—a special diode arrangement is needed. See Figure 14.8. In this situation, a pair of Zener diodes would be appropriate. Remember from Chapter 8 that a Zener diode will limit the voltage and can be used as a voltage regulator. The voltage rating of the diode would need to be only slightly higher than the source voltage so that the diode would not activate except when the motor is in a flyback condition (and a voltage spike occurs). The Zener diodes need to be placed in opposite directions so that no matter which direction the motor is turning, one of them will be available when the spike occurs.

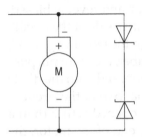

Figure 14.8: Clamping Zener diodes

To build this project, you'll need the following materials:

- Circuit created in the previous project
- Small DC motor, 6–12VDC range
- Diode, 1N4001 to 1N4007
- (2) 9V batteries with leads
- Jumpers

Use Figure 14.9 to help you build the motor circuit. The contact points referenced are for the demo breadboard and may not match your breadboard, so adjust accordingly.

1. If you *have not* created the first project in this chapter, you'll need to go back and build that circuit. If you *have* created the circuit, simply remove the two LEDs and the 220-ohm resistor.

2. Insert the anode lead of the diode in 36h and the cathode in 40h. (The anode connects to the transistor's collector.)

3. Insert the motor leads in 36j and 40j.

4. Place a jumper between 40h and 45h.

5. Place another jumper between 46h and the power rail.

6. Insert the 9V battery connectors into the breadboard, one to power the 555 timer side and the other to power the motor side.

7. Wrap a piece of electrical tape around the shaft of the motor, forming a mini flag so you can more easily see the shaft turning.

Turn the potentiometer. The motor should speed up and slow down as you turn the potentiometer in one direction and then the other.

Try This: Trying PWM and an Arduino

Now that you understand what PWM is and a bit about how it works, it's time to try PWM using an Arduino. The first mini-project simply dims the light using PWM. By adding more LEDs, the second project can be used to create a light show.

The Arduino Uno has several pins designated as PWM. They are pins 3, 5, 6, 9, 10, and 11. The PWM pins essentially replace the 555 timer circuit that was created in the previous project. The frequency of pins 3, 9, 10, and 11 is approximately 490Hz, while the frequency of pins 5 and 6 is approximately 980Hz. The duty cycle (% of time on) is determined using the `analogWrite` function. Just like with the 555 timer circuit, the output is still a digital output, but it mimics an analog output by changing the duty cycle.

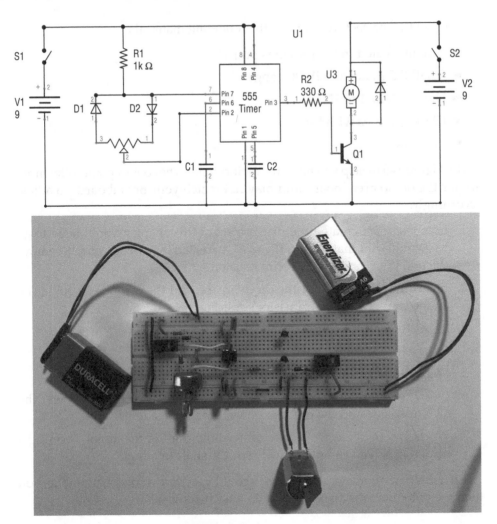

Figure 14.9: PWM motor circuit with flyback diode

The PWM pins have eight binary bits to represent the duty cycle. If all eight bits are 0, the value is 0; if all eight bits are 1, the decimal value is 255. So, 255 would represent a duty cycle of 100%, 128 would represent a duty cycle of 50%, and 26 would represent a duty cycle of 10%. Just as in the first circuit in this chapter, the percentage of duty cycle is equivalent to the percentage of brightness of the LED.

For this project, you need the following materials:

- Arduino Uno with a USB cable
- Arduino IDE installed on your PC
- LED, any color

- 220-ohm resistor
- Breadboard

It may be tempting to connect the LED directly to the Arduino's PWM pin and ground, but it is not recommended. The 220-ohm resistor is in the circuit to limit current. Use Figure 14.10 as your guide.

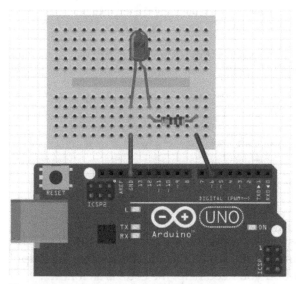

Figure 14.10: Simple Arduino PWM circuit

The circuit will start out with a 100% duty cycle using the `analogWrite` function and pause at that brightness for 1/2 second and then grow dimmer as it executes the lines of code that follow.

1. Insert the 220-ohm resistor between pins 2c and 7c.
2. Insert the long lead of the LED in pin 7d and the short lead in 9d.
3. Jumper from 9a to the Arduino's GND pin.
4. Jumper from 2a to the Arduino's pin 6.

The programming is almost as simple, but subsequent adaptations are more complex.

1. Connect the Arduino Uno to the PC using the USB cable.
2. Open the IDE, and enter the following code; remember that copy and paste works in the IDE and will save time:

```
void setup() {
  // put your setup code here, to run once:
```

```
    }

    void loop() {
      // put your main code here, to run repeatedly:

        analogWrite(6, 255); //output 100% duty cycle on pin 6
        delay(500);   //delay 1/2 second

         analogWrite(6, 191); //output 75% duty cycle on pin 6
        delay(500);   //delay 1/2 second

         analogWrite(6, 128); //output 50% duty cycle on pin 6
        delay(500);   //delay 1/2 second

         analogWrite(6, 64); //output 25% duty cycle on pin 6
        delay(500);   //delay 1/2 second

         analogWrite(6, 32); //output 13% duty cycle on pin 6
        delay(500);   //delay 1/2 second

         analogWrite(6, 0); //output 0% duty cycle on pin 6
        delay(500);   //delay 1/2 second
    }
```

The LED should get progressively dimmer; then when it's off, it will go back to full brightness. To make it dim and get brighter gradually, reverse the output values. Try this on your own, but if you need help, the code is as follows:

```
void setup() {
  // put your setup code here, to run once:

}

void loop() {
  // put your main code here, to run repeatedly:

  analogWrite(6, 255); //output 100% duty cycle on pin 6
  delay(500);  //delay 1/2 second

   analogWrite(6, 191); //output 75% duty cycle on pin 6
  delay(500);  //delay 1/2 second

   analogWrite(6, 128); //output 50% duty cycle on pin 6
  delay(500);  //delay 1/2 second

   analogWrite(6, 64); //output 25% duty cycle on pin 6
  delay(500);  //delay 1/2 second
```

```
   analogWrite(6, 32); //output 13% duty cycle on pin 6
delay(500);  //delay 1/2 second

   analogWrite(6, 0); //output 0% duty cycle on pin 6
delay(500);  //delay 1/2 second

    analogWrite(6, 32); //output 13% duty cycle on pin 6
delay(500);  //delay 1/2 second

   analogWrite(6, 64); //output 25% duty cycle on pin 6
delay(500);  //delay 1/2 second

   analogWrite(6, 128); //output 50% duty cycle on pin 6
delay(500);  //delay 1/2 second

   analogWrite(6, 191); //output 75% duty cycle on pin 6
delay(500);  //delay 1/2 second

}
```

Remember to save your work. The previous circuit dims the light, but the changes seem a bit clunky. For this circuit, the intent was to illustrate what was happening in the circuit. To have a more subtle effect, different lines of code are needed. There are several different ways to dim and brighten an LED using PWM. The following program uses do...while and if...else statements.

First, the PWM pin 6 is given the name LED1, and then a variable named v1 is established with the value of 1.

In the loop, the program is divided into two sections named grow and fade. The grow section increments the value of v1 by one at each pass and then writes the value of v1 to pin 6. It does this as long as the value of v1 is less than 255. Once it reaches 255, the goto grow; statement is ignored, and the program continues on to the fade section. It will decrement the value of v1 until v1 reaches 0, and then it will go to the grow section. The delay(10) lines can be used to increase or decrease the time it takes for the LED to gain full brightness.

Open the IDE and enter the following code (you may omit the items after the // on each line if you want, as those are comments):

```
int LED1 = 6; //names pin 6 LED1
int v1 =1; //initializes the value of v1 to one.

void setup() {
  // put your setup code here, to run once:

}
```

```
void loop() {
  // put your main code here, to run repeatedly:


  grow:
  do {  //tells the program to execute the following lines
    v1 = ++v1;  // increments the variable by 1
    analogWrite(LED1, v1); //sets the duty cycle % for pin 6 based on
v1/255
    delay(10);  //pauses 10ms

  } while(v1 <255);  //tests the conditions for the do command

  if(v1 <255) {  //loops the program back to grow until the full value
is reached
    goto grow;  // once v1 = 255, the program continues on to the fade
section
  }

  fade:
  do {    // tells the program to execute the following lines
    v1 = --v1;  //decrements the variable by 1
    analogWrite(LED1, v1);  //sets the duty cycle % for pin 6 based on
v1/255
    delay(10);    // pauses 10ms
  } while(v1 >0);  // tests the conditions for the do command

  if(v1 >0) {  // when v1 reaches 0, it continues on to the next lines
  goto fade;
  }
  else {  // since the condition in if above is false...
  goto grow;  // take this action
  }

}
```

Remember to save your work. When the program runs, the light should grow bright and then slowly dim again and repeat the process. The program could be modified to make several LEDs grow and fade alternately by assigning each a vx, where x is the number of the LED and initializing the value to a different number; for example, if there were three LEDs, v1=1, v2=128, and v3=255. Each would then be at a different brightness. They would need to be assigned a pin number at the start and a grow and fade section for each. If you need help creating the circuit or sketch, visit cliffjumpertek.com/Arduino/pwm.

There is so much more that can be done with PWM.

■ The output of the Arduino could be connected to a transistor and used to run almost any circuit or device.

- Consider connecting a photoresistor to an input pin. Then the resistance on the input pin could be converted to a value that would determine the PWM percentage to apply to the output to turn a light on at a certain brightness.

- A value retrieved from a heat sensor could be used to control the speed of a fan using PWM.

The possibilities are limited only by your imagination.

Sources of Electricity

Sunshine is a form of energy, and the winds and the tides are manifestations of energy. Do we use them? Oh, no! We burn up wood and coal... We live like squatters, not as if we owned the property."

Thomas Edison

Anyone who knows me can tell you that I've always been a "tree hugger," even when it was not popular to be. I love Mother Earth and the beauty therein. So if I seem to have a leaning toward solar power, wind power, or other replenishable sources, it's true—I do. I also know that they are not perfect.

In previous chapters, we touched on various sources of electricity, and this chapter brings it all together for you in one convenient place.

Chemical Reactions

Batteries are containers for a controlled chemical reaction designed in such a way that when a conductor path exists between the positive and negative terminals, free electrons rush from the negative terminal to unite with the holes in atoms at the positive terminal. Batteries come in many different shapes, sizes, and configurations, from the box-like deep-cycle marine batteries to the tiny, but oh-so-important, hearing aid batteries that my dad uses.

As mentioned in Chapter 7, "Series and Parallel Circuits," individual batteries can be combined into battery packs or banks, not just to get the right voltage for the simple projects in this book, but for more complicated projects such as large power storage for solar-powered homes. In Chapter 17, "Inverters and Rectifiers," we'll take a look at what's involved in a solar-powered system.

You might not be a chemist, but with a little knowledge you, too, can make a battery from various common materials. What is needed are two dissimilar elements such as copper and nickel, carbon and zinc, or graphite and aluminum, and an *electrolyte*. The electrolyte is an acid, such as lemon or grapefruit juice or salt water. Every year in my class, we have a contest to see who can achieve the greatest voltage from their battery creations, which sometimes involve several voltaic cells connected together and have been known to power lights and calculators.

The *voltaic cell* (i.e., battery) is named for its inventor, Alessandro Volta. A voltaic cell consists of two dissimilar elements placed in an electrolyte. A chemical reaction in the elements causes one element to take on a negative charge and the other to take on a positive charge. When the two are connected via a conductor, electrons will flow between them as the atoms seek to balance the charges.

WARNING   Acids can burn skin! When working with battery (or any) acid, take extra precautions to avoid getting it on your skin, in your eyes, or on other surfaces. When mixing water and acid, pour the acid into water (and not the other way around) to avoid splashes, and always wear protective gloves and goggles.

Simple Experiment: Making a Voltaic Cell

You can do this "just for fun" project in many ways, but this is likely the most simple method. See Figure 15.1.

For this experiment, you'll need the following:

- Half a lemon
- Penny
- Nickel
- Multimeter

Follow these simple steps to make electricity:

1. Cut the lemon in half. You might also want to cut a bit off the end so the lemon will sit flat. Stand the lemon on a surface with the middle cut facing up.

2. Insert the penny and nickel in the fleshy part of the freshly cut lemon, ensuring that they do not touch.

3. Set your multimeter to measure the smallest voltage measurement. You can also measure current; however, in this example, voltage rise is being measured.

4. Place the leads of the meter, with one touching the nickel and the other touching the penny. If the reading is negative, simply switch the placement of the leads.

Ta da! You just made an electromotive force (EMF)!

Figure 15.1: A homemade voltaic cell

Types of Batteries

Commercial batteries are divided into two distinct types called *primary cells* and *secondary cells*. Primary cells are those that, once expended, must be recycled or disposed of. Secondary cells can be recharged so that the chemical action can happen again. Many types of batteries exist, and Table 15.1 lists merely a few examples of each.

Table 15.1: Primary and Secondary Cells

PRIMARY CELLS	SECONDARY CELLS
Alkaline	Lead-acid (automotive)
Silver oxide	Nickel cadmium (NiCad)
Nickel-metal hydride	Nickel-metal hydride (NiMH)
Zinc-air	Zinc-air fuel cells
Mercury (no longer used)	Lithium-ion (Li-ion)
Lithium	Rechargeable alkaline
Zinc chloride	Silver-cadmium

Alkaline batteries are a type of *dry cell battery*, meaning that there is just enough moisture in the electrolyte to allow for conduction, making the electrolyte more like a paste than a liquid. Another example of a dry cell would be wetting a piece of blotter with salt water and placing it between a nickel and a penny. The salt water is the electrolyte, but there isn't enough moisture for it to drip. By contrast, *wet cells* have liquid electrolyte, and the first example that comes to mind is the acid in a car battery. Some batteries have an electrolyte in gel form.

WARNING   Bear in mind that when a battery is leaking, the substance coming out is electrolyte and most likely an acid. Handle leaky batteries cautiously. Avoid getting the substance on the skin or in the eyes.

So many battery types exist because the way they are used varies greatly. Each battery type has different characteristics that make it more or less suitable for a given purpose. Some can pack power more densely so they take up less space and weigh less than other batteries, but still provide long life between recharges. The lithium-ion batteries used in most laptops are an example. Their downside is that they can literally explode when subjected to too much heat or if they are damaged.

Some batteries like NiCad are more prone to developing "memory," which means that they lose capacity as they are charged and discharged. These batteries should be fully discharged before recharging to avoid developing memory, but others can be charged at any point without harming the battery's capacity.

With so many different types of batteries, the best course of action, whether you're choosing a power option for a circuit or purchasing a replacement battery, is to research the battery to ensure that it's the right one for your purpose and to ensure that you know how to safely care for it and extend its life as long as possible.

One thing all batteries have in common is that their capacity is measured in ampere hours (Ah) or milli-ampere hours (mAh). One Ah means that a battery can provide one ampere of current for one hour. (Ampere hours are explained more in Chapter 6, "Feel the Power.") Here are some general precautions to use with batteries, the most important being that you should always read the documentation for safe use and storage of any batteries you use:

- Do not mix old and new batteries.
- Store batteries in a cool, dry place.
- Do not used a damaged or leaking battery.
- Carefully observe polarity when installing batteries.
- Never short-circuit a battery.
- Never heat or disassemble a battery.

- Dispose of batteries properly (check with local authorities).
- Never mix primary and secondary cells.
- Always read documentation.

Try This: Making a Thermocouple

Heat by itself is a rather short topic when it comes to generating electricity. Chapter 12, "Electricity's Changing Form," mentioned thermocouples, but now we're going to create one. When two different metals are connected and heated, a current is induced in the circuit. The amount of current depends on the metals or alloys being used. Some clearly work better than others. If you would like to try this on your own, it's a fairly simple process. Our project uses copper and iron, but in commercial thermocouples many other metals or metal alloys are used. While completing this project, use caution handling the lighter, and remember that the metals will become quite hot when heated. Do not touch them or the helping hands until they've had time to cool down, and do not allow them to come into contact with any flammable materials.

For this project, you will need the following:

- Piece of copper wire
- Piece of iron or a steel nail
- Lighter with a long handle
- Helping hands
- Multimeter or galvanometer

Use the following procedure to observe the action of a thermocouple. Figure 15.2 shows an example of this process.

1. Strip a few inches off one end and 1/2 inch off the other end of the copper wire. (In the example circuit, an alligator clip was already on the other end of the copper wire, so that worked well to connect to the multimeter.)
2. Twist the longer, stripped end of the copper wire around the piece of iron.
3. Use a helping hand tool to hold the wires with the twisted section between the clamps.
4. Connect the opposite end of the copper wire to one meter probe and touch the piece of steel with the other meter probe.
5. Set the meter to measure the circuit current.
6. Use a long-handled lighter to heat the twisted joint so that your hand is farther away from the heat. Avoid melting the wire's insulation.

7. Observe the meter. The values may need to be adjusted downward as the current may be very small.

(Note: The demo circuit didn't generate enough current for the inexpensive meter to measure, but it was able to show the voltage rise in mV. Remember that when checking the voltage rise across a source, it's typically done only for a few seconds and in parallel to the circuit. However, the voltages measured here should be insignificant. A better option is to use a galvanometer if one is available. A galvanometer measures very small currents.)

8. Turn the lighter off and place it on a heat-resistant surface.

9. Allow the metals to cool down before touching them or removing them from the helping hands.

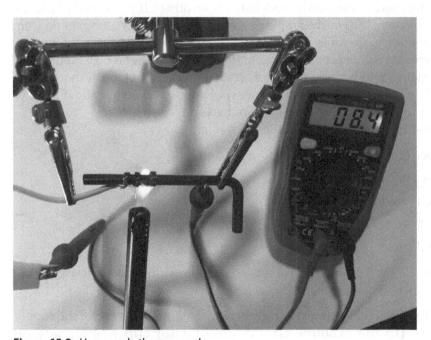

Figure 15.2: Homemade thermocouple

Light

Who doesn't love to sit in the warm sunshine coming through a window on a winter day? Heat is one way that light energy can change form; however, *photovoltaic (PV)* cells can turn light energy into electricity that can be used to support an entire home or even multiple businesses. Personally, I have a small solar-powered charger that hooks to a backpack and is used to charge my cell phone when I'm out on a hike in the mountains. PV cells come in many different forms, but how do they work?

PV cells have much in common with the bipolar junction transistors that were discussed earlier. In a simplified explanation, a PV cell is constructed of two layers of silicon. One layer is doped with phosphorous so that it is an n-type (negative) material, and the other is doped with boron, to make it a p-type (positive) material. (See Figure 15.3 for assistance in understanding the following process.) Silicon by itself is a semiconductor with four valence electrons in its outermost band. Each silicon atom shares its valence electrons with the neighboring atoms, creating what is known as a *crystalline lattice structure*. When silicon is doped with phosphorous, which has five valence electrons, one electron is unable to be shared with the neighboring silicon atoms. Conversely, when the other layer is doped with boron, wherever there is a boron atom the neighboring silicon atom will not be able to share an electron because there is a hole where the electron would be.

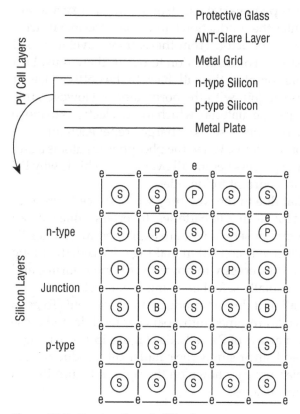

Figure 15.3: Construction of a PV cell

Where the p-type material and the n-type material are joined, the electrons from the n-type material rush to fill the holes in the p-type material. As the holes fill, more energy is required for an electron to jump the barrier they create over to the holes, and eventually the distance exceeds their energy. At this point, a potential barrier exists between the n-type and p-type material, but within the barrier the n-type material now has a positive charge where the phosphorous electrons are missing from the atoms, and the boron-doped material has a negative charge because it has acquired more electrons than it has protons.

The PV cell consists of several layers; this example is the most simplified version, as other PV cell types may have multiple layers, each absorbing a different wavelength of energy. (Refer again to Figure 15.3.) The basic PV cell will have a layer of glass to protect the fragile silicon layers, a layer of anti-glare material to keep the energy in the cell, a metal grid to collect freed electrons, then the two layers of silicon (n-type and p-type), and a metal plate on the bottom.

When light shines on the n-type material, electrons gain energy and are freed from their atoms. They enter the conduction band where the metal strips collect them and pass them to a conductor. From there, they travel to a load or are stored as energy in a battery. For conduction to occur, there must be a complete circuit. The conductors complete the path back to the bottom plate of the PV cell where the electrons can join with the boron atoms. However, this creates a negative charge in the p-type material, which forces electrons to move back through the potential barrier (because like charges repel each other) to their eagerly awaiting phosphorous atoms. When the phosphorous atoms again receive energy from sunlight, the process begins all over again. This is why PV cells are able to produce electrical current for decades.

Also important to note is that due to the way they work, PV cells produce direct current, which is perfect for storing in batteries. Inverters, which are an important part of a PV system for household use, are explained in Chapter 17.

While it's true that some pollution is created in making a PV cell, their long life and ability to use the sunlight's energy to power our homes, farms, and businesses is far more environmentally safe than using fossil fuels.

A single PV cell is quite small and doesn't produce much current. To get a significant current, several PV cells are joined into a panel or module, and those modules are joined to create vast arrays of PV cells. See Figure 15.4. The images on the left are single PV cells, and surprisingly the smaller cell produces more energy. A small section of a solar array is shown on the right of Figure 15.4. In its entirety, the array covers several acres.

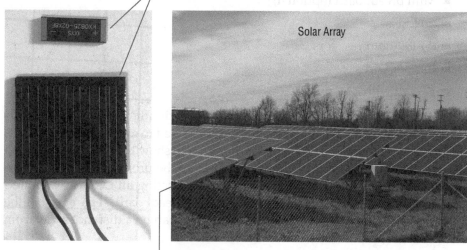

Two individual PV cells. Note the metal collector lines on the larger one.

Solar Array

A module composed of multiple cells

Figure 15.4: PV cells, panels, and array

Try This: Displaying PV Output on an Arduino

This simple project provides a visual display of the output of a PV cell. Because the PV cell used in the demo project has a maximum output of approximately 3.5V, it can be connected directly to the Arduino. If the voltage being read has the potential to go above 5VDC, then refer to the voltmeter project in Chapter 5, "Dim the Lights," to avoid damaging the Arduino board, and connect the PV cell to a voltage divider before reading the value.

Now that you have some knowledge of electronics and Arduino, this project might seem too simple. However, consider the possibility of connecting a circuit like this to Wi-Fi or the Internet, and having it report to you what's going on with your PV system. Therein lies the power of a simple system. Remember, too, that although we're using your computer to power the Arduino, in the field it would most likely be powered by a battery, or maybe even a solar cell.

This project uses an Arduino Uno, but there are smaller, less expensive Arduino boards that would work just as well.

For this project, you'll need the following materials:

- Arduino Uno and USB cable
- IDE installed on a PC

- 1/2-size breadboard
- Mini breadboard (optional)
- (5) LEDs
- (5) 100-ohm resistors
- Jumper wires
- PV cell with <5V output

See Figure 15.5 for assistance in building your circuit. Because the PV cell's output is less than 5V, it merely needs to be connected to ground (GND) and the Arduino's A0 connector. The Arduino's output will show the voltage by lighting a series of LEDs and through the serial monitor. The project could also use an LCD display for its output. (Instructions for connecting an LCD can be found in Chapter 6.) The PV cell is mounted on a mini breadboard, separate from the main circuit, to illustrate the fact that the PV cell can be a distance away from where the data is provided.

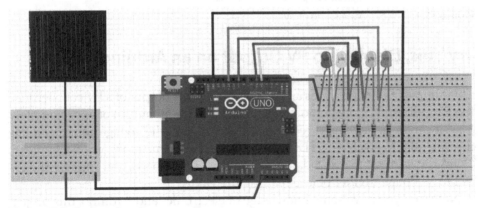

Figure 15.5: Project pinout

1. Insert the leads of the PV cell into two separate terminal rows of the mini breadboard.

2. Jumper between the PV cell's negative connector and the GND connector on the analog side of the Arduino.

3. Jumper between the PV cell's positive connector and the A0 connector on the analog side of the Arduino.

4. Insert the LEDs in a row, but not connecting, as follows:

   LED1: Anode 1c, cathode 3c

   LED2: Anode 4c, cathode 6c

   LED3: Anode 7c, cathode 9c

LED4: Anode 10c, cathode 12c

LED5: Anode 13c, cathode 15c

5. Connect a current-limiting resistor to the cathode of each LED and across the breadboard's dip by connecting a 100-ohm resistor to the points as follows:
LED1: 3f to 3g

LED2: 6f to 6g

LED3: 9f to 9g

LED4: 12f to 12g

LED5: 15f to 15g

6. Connect each resistor to ground with a jumper between the following pins:
LED1: 3k to ground rail

LED2: 6k to ground rail

LED3: 9k to ground rail

LED4: 12k to ground rail

LED5: 15k to ground rail

7. Connect each LED's anode to an Arduino PWM pin as follows:
LED1: 1a to Arduino PWM pin 3

LED2: 4a to Arduino PWM pin 5

LED3: 7a to Arduino PWM pin 6

LED4: 10a to Arduino PWM pin 9

LED5: 13a to Arduino PWM pin 11

8. Connect the ground rail to GND on the digital side of the Arduino board.

That's all for the physical connections.

Now it's time for the programming. The code will read the voltage input on the analog pin and return a value between 0 and 1,023 (1,024 possible values). That value will, in turn, be translated to one or more LEDs being fully or partially illuminated. Because there are five LEDs, each LED will be assigned a value based on 1/5 of the possible input values. For each time that the value read increments by 205, an additional light will be fully lit (1,023/5 = 204.6). If, for example, the value read on the input PV cell pin was 250, then the first LED will be lit, and the second will receive a PWM signal of 22% (250 − 205 = 45. 45/205 = 0.219). To get 22% brightness, the maximum PWM value of 255 is multiplied by the percent, so the value sent to the pin would be 56 (255 *.22 = 56.1).

For the measured voltage, be sure to change the 3.5 volts below to whatever the maximum voltage output is of the PV cell being used. This is an approximation.

For a more accurate voltage reading, or if your PV cell is capable of greater than 5VDC output, refer to the voltage divider that is used for the input of the volt-meter in Chapter 5.

The program uses the if/else functions to choose which set of output values to use, based on the value read from the PV cell's input. For example, if the input from the PV cell were 750, then the if and first else statements would both be false, but the second else statement would be true, so the actions listed under that statement would be processed.

Plug the Arduino's USB connector into the PC, launch the IDE, and enter the following code. As you enter the code, remember that copy and paste can be used to make the job easier.

```
// assigns a pin to each LED
int ledPin1 = 3;
int ledPin2 = 5;
int ledPin3 = 6;
int ledPin4 = 9;
int ledPin5 = 11;

// creates a variable for the value of each pin and sets the value of
each to zero
int value1 = 0;
int value2 = 0;
int value3 = 0;
int value4 = 0;
int value5 = 0;

//creates variables to calculate the PWM value to use
int x = 0;
long y = 0;
int z = 0;

// creates a variable to display the voltage
float volts = 0;

//creates a variable for the read value of the PV cell
int PV = 0;

void setup() {
  // sets the PV cell's pin to input
  pinMode(A0, INPUT);

  // sets the LED pins to output
  pinMode(ledPin1, OUTPUT);
  pinMode(ledPin2, OUTPUT);
  pinMode(ledPin3, OUTPUT);
  pinMode(ledPin4, OUTPUT);
```

```
    pinMode(ledPin5, OUTPUT);


    Serial.begin(9600);  // connects to the serial monitor

}

void loop() {
  // put your main code here, to run repeatedly:


//resets calculating variables to zero
int PV = 0;
int x = 0;
long y = 0;
int z = 0;
delay(10);   // added for stability

// resets the PWM output values to zero
int value1 = 0;
int value2 = 0;
int value3 = 0;
int value4 = 0;
int value5 = 0;
delay(10);   // added for stability

// reads the input value of the PV cell
PV = analogRead(A0);

// print the value read to the serial monitor
Serial.print("PV=");
Serial.println(PV);

// calculates and prints voltage
float volts= ((float)PV/1023)*3.5;   //replace 3.5 with the maximum
output of the PV cell being used.
Serial.print("Volts = ");
Serial.println(volts);

if (PV == 1023){  //If the input is at maximum, all LEDs will be fully
lit.
  value1=255;
  value2=255;
  value3=255;
  value4=255;
  value5=255;

}
else if (PV >= 818) {
```

```
    x=PV-818;                // finds the difference between the PV reading and
the sum
                             // of the fully lit light PV values for this instance
    y=x*1000L/205;           // calculates % of bright and multiplies before
dividing to
                             // avoid using a floating number
   z=round((y*255)/1000);   //Finds the PWM value for the % of brightness
;
   value1=255;
   value2=255;
   value3=255;
   value4=255;
   value5=z;

   Serial.print("x=");    // These lines are only here so you can see
   Serial.println(x);     // how the program is working and
   Serial.print("y=");    //for troubleshooting purposes.
   Serial.println(y);
   Serial.print("z=");
   Serial.println(z);

}

else if (PV >=614) {
  x=PV-614;
   y=x*1000L/205;
  z=round((y*255)/1000);
  value1=255;
  value2=255;
  value3=255;
  value4=z;
  value5=0;

}

else if (PV >=409) {
  x=PV-409;
  y=x*1000L/205;
  z=round((y*255)/1000);
  value1=255;
  value2=255;
  value3=z;
  value4=0;
  value5=0;

}
```

```
else if (PV >=205) {
  x=PV-205;
   y=x*1000L/205;
  z=round((y*255)/1000);
  value1=255;
  value2=z;
  value3=0;
  value4=0;
  value5=0;

}

else if (PV >=0) {
  x=PV;
 y=((1000*x)/205);
  z=round((y*255)/1000);
  value1=z;
  value2=0;
  value3=0;
  value4=0;
  value5=0;
  Serial.println(value1);

}
// the following writes the PWM values to the output pins.
analogWrite(ledPin1, value1);
analogWrite(ledPin2, value2);
analogWrite(ledPin3, value3);
analogWrite(ledPin4, value4);
analogWrite(ledPin5, value5);



}
```

Save your sketch. Compile and make any changes needed. Save again. Launch the serial monitor and then upload your sketch to the Arduino Uno.

Expose the PV cell gradually to light. The LEDs should light in order. In the example, they light from red to green, with green being the highest voltage. As you pass your hand over the PV cell to block the incoming light, the LEDs should grow dim and then go out. Figure 15.6 shows what the serial monitor's output should look like.

One section is highlighted for ease in determining when a single loop starts and ends. Red and blue lines indicate what part of the program caused each part of the output. Note that as the program ran, it skipped over the if statement because the PV was less than 1,023, but the next else if statement, PV > = 818, is true, so that section was executed.

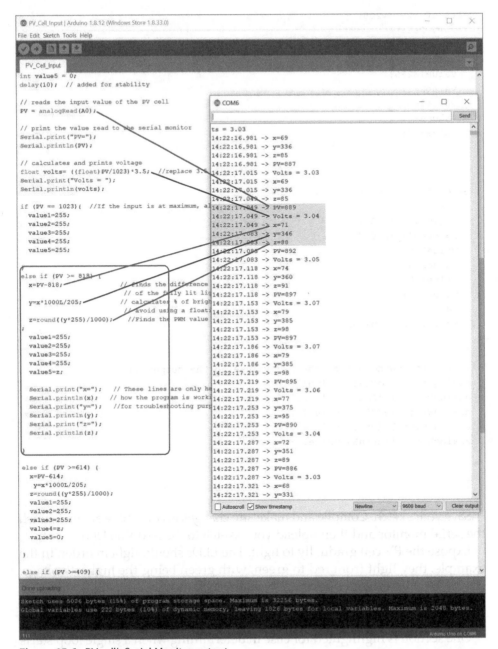

Figure 15.6: PV cell's Serial Monitor output

818 is the value that would fully light four lights, because 4/5 is 80% (0.8 * 1,023 = 818). Anything above that would partially light the next light in line, but the PWM value for that LED needs to be determined. Subtracting 818 from the total value read gives the value of that last LED, which we're calling *x*. Because one fully lit light requires an input reading of 205, *x* divided by 205 gives us the percentage of total illumination that the light should display, which we will call *y*.

In the setup section, the *y* variable is created as a long integer because it's best to avoid *floating-point* values that take a good deal of the microprocessor's memory. A floating point is any number that results in a decimal. Multiplying that value by 1,000 allows us to avoid using a floating point; the only caveat is that if the value read were significant, it would exceed the value available for use with an *int* (integer) type (+/- 32,768). So, the variable *y* needed to be created as a *long integer*, which allows values of +/- over 2 billion. If *y* were initialized as simply an integer, the result would be 0. Important to note is that not only does the *y* variable need to be initialized as long, but where it is used at least one of the values must have an L at the end to indicate that it is a long integer.

Finally, the PWM value needed to achieve the correct illumination can be found by multiplying the percentage of illumination by 255, which is the value for 100% PWM. (PWM is explained in Chapter 14, "Pulse Width Modulation.") The value must be divided by 1,000 to correct for multiplying by 1,000 in the preceding line.

Once the section of code is executed, the program will skip the remaining `else if` statements, begin processing again at the `analogWrite` statements, and then skip back up to `void loop` and start over.

Figure 15.7 shows the circuit in action.

Friction

You may be asking yourself, "Friction as a source of electricity?" Yes. It can be, and not just the static jolt you get when touching a doorknob after walking across a carpet. Creating electrical current by friction is known as the *Triboelectric effect*. The people at Disney Research in conjunction with Carnegie Mellon University have been researching using the triboelectric effect to create paper generators that can be used for interactive books for children that light up and produce motion from tapping or rubbing paper and Teflon together. The charges generated can be enough to power an ebook reader.

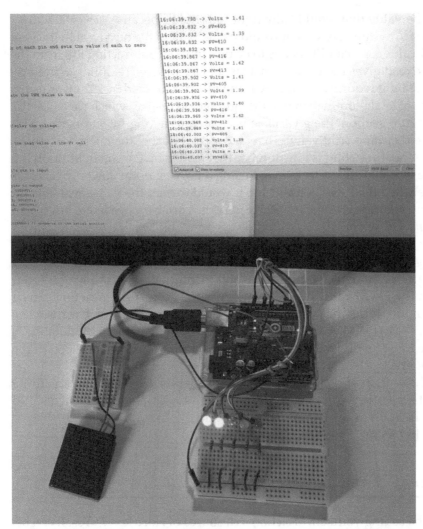

Figure 15.7: Completed PV monitoring circuit

Another team from the University at Buffalo and Kansas State University are studying how the triboelectric effect could be used to power everyday electronics like a smartwatch or smartphone. Keep watching for them because paper generators and other triboelectric devices are on the way.

Magnetism

Magnetism is the driving force behind much of our power generation. When magnetism is used for power generation, some sort of mechanical force is needed to push against the blades of a turbine, which in turn rotates a shaft within a magnetic field. The source of that mechanical force is commonly the flow of water, like at the Hoover Dam on the border between Arizona and Nevada, or the force of wind in the wind farms popping up all over the world, but it can also be the pressure of steam from burning some other material like wood or fossil fuels. In fact, nuclear power uses chemical reactions to generate heat and convert water to steam, which then turns large turbines and creates electrical current. Even solar or geothermal energy can be used to move turbines.

The oceans are a great source of energy in the form of waves, and while a practical solution to harnessing the ocean's power isn't a reality yet, there are several companies working on solving the problems. At the time of this writing, wave energy is being researched and implemented in several parts of the world including Scotland, the Caribbean, the United States, and Australia. One of the problems faced in this growing industry is finding equipment sturdy enough to withstand the ocean's wrath and then converting the sporadic wave movements into something reliable enough to move the turbines of a generator. According to the U.S. Energy Information Administration website, there is enough energy in ocean waves to provide more than 60% of the United States' power needs.

Whatever the source of the mechanical power is, the shaft of the turbine is wrapped with multiple conductors that are connected to external conductors via brushes or slip rings. The shaft is surrounded by magnets or electromagnets that provide the magnetic field, and as we learned in Chapter 11, "The Magic of Magnetism," whenever there is a conductor, a magnetic field, and relative movement between them, a current is generated. Figure 15.8 shows a dissected motor. This is a small 6VDC motor, but the magnets and coil surrounding the shaft are visible. As mentioned in Chapter 11, motors and generators have the same basic construction, to the point that if something were attached to a motor's shaft to turn it, the motor could become a generator.

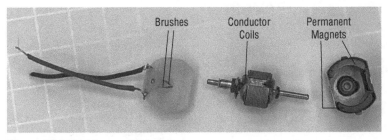

Figure 15.8: Inside a motor/generator

Pressure

A final and sometimes overlooked source of electricity is pressure. Chapter 12 introduced you to piezoelectrics (pronounced pee-ay-zo-electrics.) Certain crystals, ceramics, fiber optics, and some organic materials like bone and even DNA will produce a current when compressed. These same materials will respond with movement when a charge is applied to them.

Piezoelectric materials are used to power devices such as the ignition systems on grills, lighters, and gas stoves. They are also popular as sensors because they can react to such a small change in pressure, even from a sound wave. They're used in bridges and large buildings to monitor structural integrity, and even to provide power to pacemakers without having to perform surgery to replace batteries.

If you would like to try making electrical current using piezoelectric materials, look for more information on `cliffjumpertek.com/piezo`.

Wrapping It Up

In summary, the sources of electrical current, at least the ones we know about now, are chemical reactions, heat, light, friction, pressure, and magnetism. Making electricity is fun and likely easier than most people think.

In the next chapter we'll examine how electricity's characteristics are changed so that it can be transmitted over long distances. See you there!

CHAPTER 16

Transformers and Power Distribution

"If history is any indication, all truths will eventually turn out to be false."

—Dean Kamen

One thing about the world that is always the same is that there will always be change. The sooner we learn that and learn to roll with the changes, the easier life will be, because fighting change is a futile battle.

In electronics, we have devices called *transformers* that help us to change the characteristics of electricity, so we can accomplish amazing feats with it. Read on to learn how transformers allow us to intentionally cause change and work it to our advantage.

What Is a Transformer?

You use transformers every day, although you might not realize that you do. Figure 16.1 shows a few common transformers. The largest one shown here on a power pole is for controlling the power sent over long distances, which is explained more toward the end of this chapter. The ever-present USB power blocks reduce the 120VAC in a wall outlet down to the 5VDC needed for our personal electronic devices, and yet other transformers, such as the tiny one shown in Figure 16.1, are used for transmitting audio signals. Transformers are the heart of power supplies. Many transformers of various sizes and configurations are used in industry to provide the myriad of voltages needed for industrial equipment, including the 3.3V, 5V, and 12V needed for a computer to work properly.

Figure 16.1: Common transformers

A transformer uses the power of magnetism to transfer an electromotive force (EMF) from one circuit to another. The total power in the circuit doesn't change. Either the circuit's characteristics change, or the transformer simply passes the power from one circuit to another while keeping them isolated from each other. Transformers (with one exception) are formed by two or more coils of conductors that are typically not physically connected but are connected through a moving magnetic field. As learned in previous chapters, whenever there is a magnetic field, a conductor, and relative movement between the two, a current will be induced. When this is done through two or more coils of conductors connected through a magnetic field, it is called *mutual induction* or *transformer action*.

Transformers require alternating current, because without the rising and falling magnetic field there would be no movement. Direct current simply won't work to transfer the energy on a continuous basis. Figure 16.2 depicts a typical transformer with its wire coils wrapped around a common metal core. In Figure 16.2, a single piece of metal core lies next to the transformer. The metal pieces are laminated together, the primary and secondary wires are wrapped around the center, and then the bottom pieces are laminated and attached (more about this later). The coil on the input side is called the *primary*, and the coil on the load side is called *secondary*. The primary side will typically have two wires extending from the coil—in the case of this transformer, the primary wires are the white-and-black ones on the left. On the medium-sized transformer in Figure 16.1, the primary side wires are the two red wires toward the top of the image. The secondary side may have two, three, or more wires.

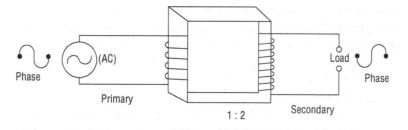

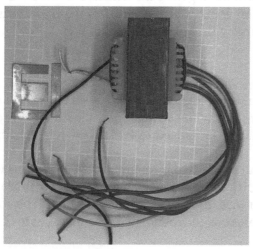

Figure 16.2: Inside a transformer

As the voltage on the primary side rises to a peak, a magnetic field is created around the primary coil. When the rising magnetic field cuts across the conductors of the secondary side of the transformer, a current is induced in the secondary side. Because the input is alternating current (AC), the voltage falls back to zero, and the magnetic field collapses. Then the field goes to its most negative point, which is sometimes called a *trough* or simply the negative peak, and induces a current in the secondary coil that flows in the opposite direction of the first.

Sometimes, the cores of a transformer are metal, like the one shown in Figure 6.2. However, radio frequency applications often use an air core transformer because less power would be lost with an air core than with a metal core. The faster frequency of a radio signal means that the polarity would change more rapidly, and as the magnetic fields collapse and rise, that changing polarity creates more molecular friction and heat loss. The actual core of a so-called air core transformer may be paper or some other nonferrous material simply used to support the coils, whereas a metal core is used to enhance the magnetic field.

Often the coils of a transformer are wrapped around the same metal core, like the center section of the single metal piece shown at left in Figure 16.2. Wrapping both the primary and secondary coils around a single core increases the number of wire coils cutting the lines of flux of the magnetic field. The wire used in these transformers is coated with a very thin insulating material, such as enamel, lacquer, or shellac, and is called *magnet wire*. The wire in the coils must be insulated, because otherwise it would short out against itself and not create the desired magnetic field. Lacquer is often used for insulation because it is thin enough that it allows for maximum transfer of power between the primary and secondary sides.

The magnetic field can be enhanced by a metal core, greater current, more turns of wire, how tightly wound together the wire is, and the particular wrapping pattern used. The number of turns of wire affect the characteristics of the power being transferred and will be explained soon.

Phase Relationships

If the voltage for each side were plotted, we could observe that the waves created are 180 degrees *out-of-phase* with each other, as shown in Figure 16.2's top image. (*Phase* is a mathematical term used to explain the relationship between two points.)

Please now refer to Figure 16.3. Two measurements are *in-phase* when their instantaneous values have the same polarity, and they cross the zero axis at the same point in time. Otherwise, the waves are out-of-phase. The condition of being in-phase or out-of-phase is independent of the magnitude of the waves being compared.

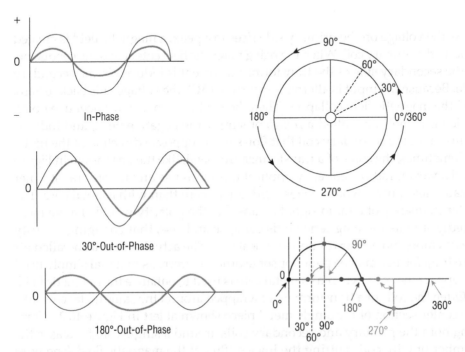

Figure 16.3: Phase relationships

Perhaps the easiest way to understand the degrees of phase difference is to consider a compass and a generator. A compass is round and is divided into 360 degrees. A generator turns in a circle within a magnetic field, with one complete rotation being equal to 360 degrees. At one point in its rotation, the generator will be 90 degrees from where it started. If we were to plot the voltage at that point, it should be at the peak voltage, right before the magnetic field starts to collapse. Examining the angle created between the instantaneous voltage at that point and the 0 axis, we will see a 90-degree angle. As the generator continues to turn, it reaches 180 degrees, which is back at the zero point on the axis, then 270 degrees at the most negative point, and finally back to 360 degrees. These degrees of rotation between the starting point and the instantaneous measurement are the phase angle used to describe an AC wave.

In examining the wave that is 30 degrees out-of-phase in Figure 16.3, the red (heavier) line is said to be *lagging* the blue line by 30 degrees, because it crosses the 0 line 30 degrees after the blue line does. It is also referred to as being negative 30 degrees out-of-phase with the blue line because it is still negative when the blue line crosses into positive. Conversely, we can say that the blue line is *leading* the red line by 30 degrees because it is positive and crossing the 0 line at a point 30 degrees before the red line does. Leading and lagging by x degrees is often how the phase relationship of two waves is described. The Greek letter phi, Φ, is used to represent the phase angle.

Power and Turns Ratio

If the coils on the primary side and secondary side of a transformer are identical, then the voltage and current transferred to the secondary side will be the same as the primary side but flowing in the opposite direction at any given instant. They are 180 degrees out-of-phase. The circuits are not physically connected, but because transformers have a very high efficiency, very little energy is lost. Transformer losses will be explained soon.

If, however, there are an unequal number of turns of wire from one side to the other, the transformer changes the characteristics of the secondary side circuit. In this regard, transformers will be one of three types: isolating, step-up, or step-down. (The isolating type was just described.) To recap, with an isolating transformer, there is no change of voltage or current from one side to the other, and the transformer is used merely to electrically isolate one circuit from the other, while passing the power of the first circuit onto the second.

The top of Figure 16.2 depicts a step-up transformer. When there are more turns of wire on the secondary side than the primary side, the voltage on the secondary side is increased by the ratio of the wire turns. Step-up and step-down refer to what happens to the voltage from one side to the other, not the current. If there are more turns of wire in the primary side than the secondary side, then the transformer is a step-down transformer. If there are more turns of wire in the secondary side, then the transformer is a step-up transformer.

The comparison of the turns of wire from one side to the other of a transformer is called the *turns ratio* and is stated with the primary side first, usually with one number or the other being a 1. For example, if there were 1,000 turns of wire on the primary side and 2,000 turns of wire on the secondary side, the turns ratio would be 1:2. This would be a step-up transformer.

Supposing that the voltage on the primary side of that 1:2 transformer was 12V, then on the secondary side of the transformer the voltage induced would be 24V. The current would have the opposite effect. So, if the current on the primary side of this transformer were 1A, the secondary side would have only 0.5A of current. Applying Watt's law to both sides of the transformer illustrates that the power in the transformer will remain the same. P = IE, P_P = 1(12), P_S = 0.5(24), both of which equal 12 watts of power.

If the transformer were a step-down transformer with a ratio of 10:1, then there would be 10 turns of wire on the primary side for each turn of wire on the secondary side. A voltage input of 120 would result in an output of 12VAC. If I know the power on the primary side is 12 watts, I can assume the power on the secondary side is 12 watts as well. I can use that figure and Watt's law to determine the current on each side. The relationship between voltage, current, and the turns ratio can be summarized as follows, where X_P is the primary side, X_S is the secondary side, and *N* is the number of turns of wire:

$$\frac{N_P}{N_S} = \frac{V_P}{V_S} = \frac{I_S}{I_P}$$

Examine Figure 16.4 for some examples of this relationship. Figure 16.4 also shows the symbols for an air core and iron core transformer.

Transformer Losses

While transformers tend to be extremely efficient electronic devices, there are some losses that occur within them. They are called copper, or I^2R (I squared R), losses; hysteresis; and eddy currents.

Copper losses are the energy lost as heat due to the resistance of the copper conductors in a transformer. Remember the power wheel from Chapter 6, "Feel the Power"? One of the formulas for power was I^2R, so the current squared multiplied by the resistance of the conductors equals the power lost due to that resistance.

Hysteresis is another type of heat loss, but this loss is due to the molecular friction when the polarity of magnetic particles changes direction, as it must whenever transformer action is involved.

Eddy currents are whirlpools of electricity that tend to form in the conductors, again due to magnetic induction. Remember the laminated core in Figure 16.2?

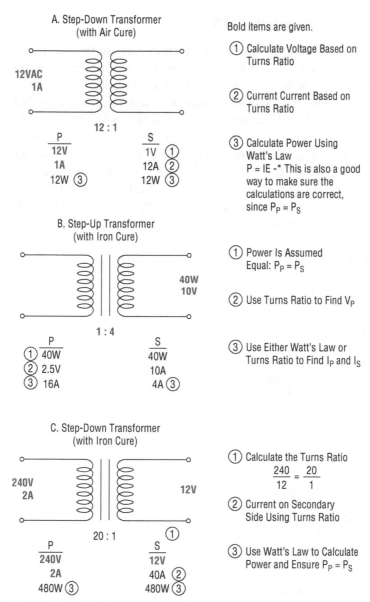

A. Step-Down Transformer
(with Air Cure)

12VAC
1A

12 : 1

$\dfrac{P}{12V}$ $\dfrac{S}{1V}$ ①
1A 12A ②
12W ③ 12W ③

B. Step-Up Transformer
(with Iron Cure)

40W
10V

1 : 4

$\dfrac{P}{①~40W}$ $\dfrac{S}{40W}$
② 2.5V 10A
③ 16A 4A ③

C. Step-Down Transformer
(with Iron Cure)

240V
2A

12V

20 : 1 ①

$\dfrac{P}{240V}$ $\dfrac{S}{12V}$
2A 40A ②
480W ③ 480W ③

Bold items are given.

① Calculate Voltage Based on Turns Ratio

② Current Current Based on Turns Ratio

③ Calculate Power Using Watt's Law
P = IE -* This is also a good way to make sure the calculations are correct, since $P_P = P_S$

① Power Is Assumed Equal: $P_P = P_S$

② Use Turns Ratio to Find V_P

③ Use Either Watt's Law or Turns Ratio to Find I_P and I_S

① Calculate the Turns Ratio
$\dfrac{240}{12} = \dfrac{20}{1}$

② Current on Secondary Side Using Turns Ratio

③ Use Watt's Law to Calculate Power and Ensure $P_P = P_S$

Figure 16.4: Calculating current, voltage, and turns ratios

The core is laminated and pieced together, rather than being a single solid core, to help reduce the amount of energy lost to eddy currents.

Choice and design of materials for the transformer core can mitigate these losses, which tend to be small in transformers anyway. These three types of losses can also occur in motors and generators because of how they are constructed.

Taps, Autotransformers, and Variacs

Autotransformers and variacs don't follow the normal rules of transformers. Instead of having two coils, they have only one, so there is no isolation between the primary and secondary coils. Figure 16.5 shows a schematic for a variac, which is a type of autotransformer. The advantage of an autotransformer is that it's less expensive to manufacture because of its construction with only one coil. If the auto transformer is a variac, then the voltage ratio can be selected, usually by turning a dial, which connects one end of the secondary output to a *tap*. Multiple taps mean that there are different voltage choices that can be made. For example, perhaps connecting to tap A would give a 1:1 ratio, but tap B would give a 1:3 ratio. If the autotransformer is not a variac, then the secondary side is fixed, but there is still only one coil used.

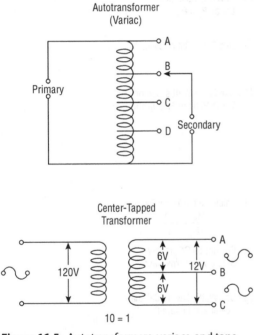

Figure 16.5: Autotransformers, variacs, and taps

Variacs are not the only transformers with taps. Often a transformer will have a center tap that divides the output voltage into two, but I have seen as many as seven taps on a transformer to provide different voltages that are needed. The taps are not always symmetrical. For example, a transformer used in a PC power supply would need to provide 3.3V, 5V, and 12V.

For the center-tapped transformer in Figure 16.5, if a connection were made between points A and B, the result would be 6V. If a connection were made between points B and C, the result would be 6V. But if a connection were made between points A and C, the result would be 12V. Most transformers have a label explaining the input and output voltage configuration. But as with all other electronics, the best source of information is often the component's datasheet.

Multiple winding transformers can have multiple secondary windings, where each secondary winding can be either step up or step down independently. Some transformers may also have multiple windings on the primary side.

The label shown in Figure 16.6 is for the transformer that is dissected in Figure 16.2. Notice how this transformer appears to have three separate windings, two of which are tapped and one which is not. There could be a total of five different circuits connected to this transformer, and the voltages are indicated by the color of the wire—9V, 16V, and 6V. Notice also that the current is indicated for each winding. Primary (Pri.) and secondary (Sec.) sides are clearly labeled. This is obviously a transformer whose primary side is intended to be connected to standard U.S. household power.

Figure 16.6: A transformer label

WARNING   Alternating current is more dangerous and deadly than direct current. Always use caution when working with transformers.

Try This: Verifying Transformer Output

The circuit in this lab will demonstrate the power of a transformer, without you having to connect it to household current, for safety reasons. It is not intended to be permanently connected and should be used only to observe and verify how a transformer changes the voltage of a circuit and transfers energy from one circuit to another.

Transformers can work using alternating current only, and because we're not using household AC, the first step is to build an oscillator. An oscillator can be constructed in many ways, such as using a relay or a multivibrator circuit with multiple transistors. The method chosen for this oscillator is a circuit using the familiar 555 timer chip.

WARNING   No electronic component is perfect. This circuit is for demonstration purposes only, not intended for a permanent connection. If any component on any project feels or smells hot, disconnect power immediately.

Figure 16.7 may help in assembling the first part of the circuit. First, build the timer circuit and verify that it works; then, continue with the transistor and transformer portions of the circuit.

For this project, you need the following materials:

- (2) half-size breadboards
- 555 timer IC
- (2) 1k-ohm resistors
- 2.2k-ohm resistor
- 22k-ohm resistor
- 0.47 uF capacitor
- 0.01 uF capacitor
- TIP41C NPN power transistor
- (2) LEDs, any color
- Assorted jumper wires
- Step-up transformer, 12.6VAC to 110VAC

The following instructions reference the positions on the breadboard used in the demo circuit. The breadboard you use may be slightly different, so adjust the reference points as needed.

1. Jumper from the positive rail on one side to the positive rail on the other side, and from the negative rail to the negative rail, so that positive power and ground can be reached from either side of the board.

2. Insert a 555 timer IC across the dip such that pin 1 is in position 10e and pin 8 is in position 10f.

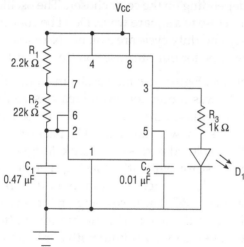

Figure 16.7: A 555 timer circuit

3. Insert a jumper between each of the following points:

    10j and the positive power rail

    10a and the ground rail

    13a and the positive power rail

    11d to 12g, connecting pins 2 and 6

4. Insert a 2k-ohm resistor between 11j and the positive power rail.

5. Insert a 22k-ohm resistor between 12h and 11h, connecting pin 6 to pin 7.

6. Insert one lead of the 0.01uF capacitor in 13j and the other lead in the ground rail.

7. Insert one lead of the 0.47 uF capacitor in 11a and the other lead in the ground rail.

8. Insert one lead of a 1k-ohm resistor in 12c and the other in 16c.

9. Insert the positive lead of an LED in 16a and the negative lead in the ground rail.

10. Strip all the insulation off a 3/4-inch piece of 22-gauge wire; then bend and insert both ends anywhere in the ground rail to become a test point.

11. Connect the positive and negative leads from a 9V battery into the appropriate power rails.

The LED should be on and appear to be constant, but it is not. Connect an oscilloscope across the two legs of the LED, and it should show a frequency of approximately 60 Hz and a duty cycle around 50%. The voltage drop across the LED will vary, depending on the color chosen. The oscilloscope should reveal a signal that's close to a square wave. Don't be concerned at the moment if the frequency and duty cycle are a bit off. No component is perfect, and it doesn't have to be for our purposes on this project.

When you're ready, continue. Figure 16.8 may assist in connecting the transistor section. The TIP41C was chosen for several reasons. At the time of this writing, they can be purchased from Digikey.com for less than $1 U.S., but more importantly, it is a power transistor. The package it is in is called a TO-220, and the silver-colored part is a heat sink designed to help draw heat away from the transistor while it's working. The hole in it is for connecting a larger heat sink if needed. The pinout of this transistor is also different—BCE instead of EBC. The base being on the outside is to our advantage when hooking up the wires. The maximum emitter-base voltage is 5V, which makes it perfect for working with a microprocessor like an Arduino. The maximum collector-emitter voltage is 100V, and the collector maximum current is 6A sustained, with a peak of 10A. This transistor can dissipate a whopping 65W of power. A regular switching transistor like the small ones that we have been using would likely overheat, causing the plastic casing to start bubbling. This transistor may get hot, but it is designed to handle that safely. All of the relevant information is available on the transistor's datasheet.

12. Disconnect the power from the circuit.

13. Remove the LED.

14. Insert the NPN transistor so that the base is in 19e, the collector is in 20e, and the emitter is in 21e.

15. Jumper between 16a and 19a, connecting the resistor to the base of the transistor. The signal from the 555 timer will switch on the transistor when the output is high.

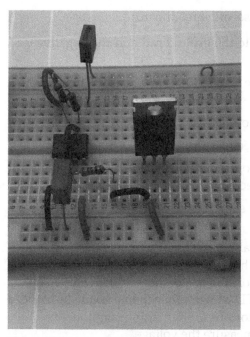

Figure 16.8: A transistor section

16. Jumper between 20a and the power rail, connecting the collector to positive power.

Examine your transformer. The transformer that we're using in the demo is intended to be a step-down transformer, from 110V to 12.6V. However, we're going to turn it around and connect it in the opposite direction so that it will become a step-up transformer instead. The 12.6V side has three wires: two yellow and one blue. The blue wire is a center tap, and if the transformer were hooked up as intended, a connection between the center tap and either yellow wire would yield 6.3V. Dividing 110 by 12.6 tells us that the turns ratio of this transformer should be 1:8.73. That means that if the input to the transformer is 2V, the output would be 2*8.73, or 17.46V. Now it's time to connect the transformer to the circuit. Remember to connect the side with fewer turns to the circuit so that the voltage will be stepped up. Figure 16.9 may assist you.

17. Insert one outside (yellow) wire of the transformer into 21a, which will connect it with the transformer's emitter. The bare wire should be long enough so that a small portion of it shows above the breadboard, and the oscilloscope can be connected at that point.

18. Insert the other outside wire into the ground rail.

19. Insert the center tap into any UNUSED terminal row, such as position 30h, to keep it out of the way and to avoid making an unintended connection.

Perform the following steps on the second breadboard:

1. Connect an LED's positive lead to the ground rail and the negative lead to 7a.

2. Connect a second LED's negative lead to the ground rail and the positive lead to 10a.

3. Jumper between 10b and 7b to connect the two LEDs.

4. Connect a 1k-ohm resistor between 7d and 2d.

5. Connect one of the red wires from the transformer to the same ground rail that the two LEDs are connected to.

6. Connect the other wire from the same side of the transformer to 2a, completing the circuit.

   The reason for two LEDs is that, although it may look like they are continuously on, in reality they are each on only about half of the time. The two LEDs are facing opposite directions, so one of them will always complete the circuit. Without the second LED, a huge voltage spike would be induced in the circuit, and spikes are notorious for damaging components. Get your oscilloscope ready to measure the voltage.

7. Connect the power to the 555 timer circuit via the power rails. The LEDs on the other breadboard should light. If they don't, disconnect the power and check circuit connections.

Figure 16.9: Completed transformer circuit

Now it's time to test the circuit.

1. Disconnect the power.

2. With the power disconnected, connect the black lead (ground) of your oscilloscope to the negative test point you created earlier.

3. Connect the red (positive) oscilloscope lead to the yellow transformer wire.

4. Connect the power to the 555 timer circuit via the power rails, being careful not to touch the secondary side of the transformer.

5. Observe the oscilloscope. There should be a somewhat square wave, although you'll likely see a spike and dip as the current switches between off and on. Jot down the frequency, duty cycle, and V_{max}, or take a picture of your oscilloscope screen to capture the values.

WARNING It's important to disconnect the power here before touching the circuit, because the voltage will be higher on the secondary side of the transformer.

6. *Disconnect the power and then the oscilloscope leads.*

7. Connect the oscilloscope leads to the other breadboard (secondary side) so that the resistor and one LED are between the two leads. Polarity doesn't matter because this is AC, and the polarity will constantly change. If you happen to have a dual-trace oscilloscope, both sides of the transformer could be connected at the same time, so you could observe the waves and their values simultaneously.

8. Connect the power again to the 555 timer circuit.

9. Observe the frequency and duty cycle on the secondary side of the circuit. They should be the same as the primary side or very close. Unless your oscilloscope auto adjusts, the wave form should look very different because the voltage is higher. Use the features of your oscilloscope to adjust the settings until the square wave appears again. The wave should look nearly identical to the wave form created on the primary side of the transformer. Figure 16.10 shows the measurements of the demo circuit.

Notice that on the secondary side of the demo circuit, the volts per division was changed to 10V so the waveform would fit on the screen, and that the waveforms are almost identical. The frequency and duty cycle show as slightly lower on the secondary side (they should be the same); however, the voltage on the secondary side is clearly higher. V_{rms} on the primary side shows 1.7V, and on the secondary side we would expect over 14V (1.7*8.73). However, it's only about 10V, so we know there are inefficiencies in the circuit.

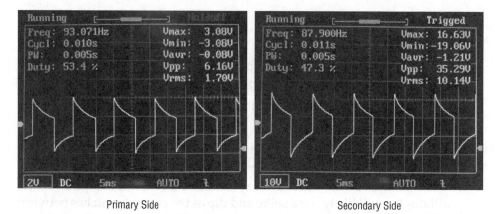

Primary Side   Secondary Side

Figure 16.10: Demo circuit readings

Notice the V_{max} on the secondary side. A component designed for a maximum voltage of 10 might break down at the peak voltage of 16.6. However, most components can handle a higher voltage for a brief time. Again, this is where the datasheet must be consulted.

How much did the V_{max} in your circuit increase by? It should have been substantial. If the primary side voltage is multiplied by the higher number in the turns ratio, is the result more than the voltage on the secondary side? It should be close, but if it isn't, it would indicate an inefficiency in either the transformer or the circuit.

The duty cycle on both sides should be the same or close. The frequency should also be the same on both sides of the transformer.

Alternating Current Values

Your oscilloscope might also have the following values identified:

- V_{max} would be the highest point that the sine wave reaches—the peak voltage. This is also known as the *amplitude* of the wave.

- V_{min} will be a negative number. This is the most negative point of the wave, sometimes called the negative peak or the trough.

- V_{pp} is the voltage peak to peak—the total difference between the V_{max} and V_{min}. If the Vmax were 10V and the V_{min} were 10V, then the V_{pp} would be 20V.

- V_{rms} stands for voltage root-mean-square. It is a formula used to find the DC equivalent of an AC wave. It is also called the *effective value*, because it's effectively the same voltage as DC. It's not the average, because the average of all the points of an AC wave would be zero. V_{rms} is often referred to in AC circuits, and it may say V_{rms} on your multimeter, too.

When determining what components to choose, it is important to be aware of the maximum sustained and instantaneous voltages and the current that a component can handle. The following formulas might help you in determining whether a component is suitable. These same formulas work for current in an AC circuit, and they work for the negative side of the wave as well as the positive side. Remember that V_{max} is the peak voltage or the amplitude of the wave, and V_{rms} is the effective value; you may see it written as E_{eff} for effective EMF, which is measured in volts.

$$V_{max} = 1.414 \times V_{rms}$$

$$V_{rms} = 0.707 \times V_{max}$$

There is an average value that is also sometimes used with circuits, so the formulas are included here, although the RMS values are the ones you'll see more often. Because the average of a complete sinewave would be zero, the average used is for half of a wave.

$$V_{max} = 1.57 \times V_{avg}$$

$$V_{avg} = 0.637 \times V_{max}$$

If you memorize only one set of these equations, remember the RMS values. Other values in AC circuits are calculated differently than DC circuits and are beyond the scope of this book.

Power Distribution Using Transformers

One final note about transformers is how they are used in power distribution. Remember the power wheel and I^2R? Resistance and current are both factors in power loss when energy is transmitted over a conductor. However, the effect of current on loss grows exponentially (it is squared), rather than linearly. For example, if power is being transmitted over a conductor with a resistance of 1k ohms, 1,000V, and 1A, then the power according to Watt's law is IE, or 1,000W. The power loss due to the resistance of the conductor is I^2R, which in this case is also 1,000W.

If a transformer is used to decrease the voltage to 100V, the current would become 10A, because $P_P = P_S$. The resistance does not change, so the power loss due to "copper" losses is now $10^2*1,000$, which would be 100,000 watts. If the current were decreased instead of increased, then the same effect would happen but in reverse. With the same resistance of 1,000 ohms, suppose a transformer is used to increase the voltage from 1,000V to 50,000V. The current on the secondary would decrease by the inverse of that ratio and would be 1/50 or 0.02A. (0.02*50,000 = 1,000W, using $P_P = P_S$ to prove the math.) Now the copper losses are I^2R, or $0.02^2*1,000$, or a total of 0.4W.

Power companies use transformers to boost the voltage up very high, so their copper losses will be minimal when sending power over a distance. This means the power companies are able to transfer more power to homes and businesses at a lower cost, and it's good for the environment because less energy is wasted. The voltages between substations may be 130,000V or more. Local substations use transformers to raise the current and lower the voltage to send power out to communities, where transformers on power poles outside U.S. homes reduce the voltage again to the 120VAC used in homes.

Once in our homes, we again use a transformer and rectifier circuit (explained in the next chapter) to drop and convert the voltage to the 5VDC that is used to charge our everyday electronics. If you're using a laptop right now with a "block" in the wire between the laptop and the wall outlet, that block contains a transformer to drop the voltage down from 120VAC to a lower voltage, and a rectifier to make it DC, changing not only the voltage but the type of current. The next chapter will explain how this is done.

CHAPTER 17

Inverters and Rectifiers

"There are a million cheap seats in the world today filled with people who will never be brave with their own lives, but who will spend every ounce of energy they have hurling advice and judgment at those of us who are trying to dare greatly... If you're criticizing from a place where you're not also putting yourself on the line, I'm not interested in what you have to say."

—Brené Brown

Dare greatly. Ignore people speaking from the cheap seats. You're about to make a power supply.

Whenever I introduce this topic to a group of high school kids, I have to wait for the giggles to die down before I can start. The word sounds funny, but rectifiers are something you use unaware every day, and perhaps inverters, too. Rectifiers and inverters are like two sides of a coin—they both perform useful functions, but what are they?

Inverters vs. Rectifiers and Their Uses

If you did the lab in Chapter 16, "Transformers and Power Distribution," you actually created a crude inverter circuit. An *inverter* takes a DC signal and inverts it into an AC signal. Remember, not all AC signals are sine waves. They can be square waves and sawtooth waves as well.

A *rectifier*, on the other hand, takes an AC wave and converts it to a DC wave, smoothing out all those ups and downs. First, we'll take a look at inverters and how they're used in photovoltaic (PV) systems, i.e., all those solar panels you see on the homes of tree huggers like me.

Inverters and PV Systems

Anyone with a PV system supplying the power to their home should be able to tell you all about their inverter. The output of a PV array is DC, so unless a homeowner with a PV system has purchased all DC appliances and lights, they're going to have an inverter to turn that signal into AC. While it is possible to get things like a DC refrigerator or blow dryer, they're typically much more expensive and less common, which makes getting parts and service more difficult. Figure 17.1 shows where an inverter fits into a DC system.

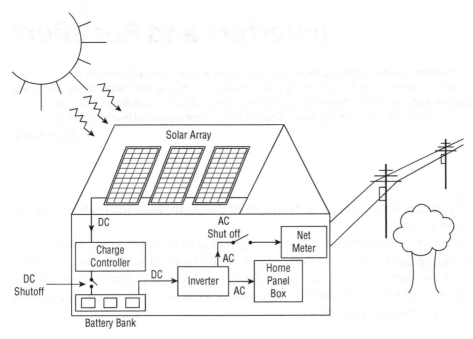

Simplified Residential PV System

Figure 17.1: Typical PV system

Somewhere between the solar panels and the battery bank of a PV system will be a shutoff and likely a safety ground. Another shutoff will be located between the inverter and the grid, so that if the grid power is down, someone working on the grid's conductors won't be electrocuted by power entering the grid from the home PV system. If the system is purely *net metered*, meaning that the electric company will either pay or charge the customer depending on whether they've contributed to the grid system or drawn more power from it than they contributed, then it may not have a battery bank. However, many PV systems do, so that if the grid power goes down, they can draw from their charged batteries. The batteries used may be deep-cycle marine batteries.

A PV system will most likely have a charge controller, which may also be the DC shutoff. The charge controller ensures that the battery bank is not overcharged and limits the current to an acceptable amount. The charge controller regulates both the voltage and current going to the rest of the system from the PV cells.

PV systems will likely have an inverter whether or not they are grid-connected. But if grid-connected, they must have an inverter, and power from the inverter will be sent through a net meter to the power grid. It will also traverse a transformer to match up with the grid's power requirements. As discussed in Chapter 16, mass power is sent over the power lines as an AC signal, and transformers are used to boost the voltage, so there is less power loss due to resistance in the conductors over those long distances.

Distributed power systems refer to the making of power locally, as in the systems used by individuals to harvest renewable energy from the wind, water, and sun. These systems are called distributed power because rather than the power being generated all in one centralized place, such as the power plant, it could be generated in a thousand grid-tied systems owned by you and your neighbors, all of whom sell their excess energy to the grid and are able to purchase energy from the grid when they need it.

PV systems can be stand-alone or grid-tied systems, and that may depend on how far off the beaten path those systems are. PV systems are also available for camper trailers, which largely run on 12VDC. In fact, you might see a camper trailer with a few PV modules mounted on its roof. I know I have.

So much more can be learned about PV systems, and while the tree hugger in me wants to explain all about PV home systems to you, that's a story for another book.

Other Inverter Uses

Car batteries and deep-cycle marine batteries (like those often found in PV systems) provide 12VDC. Figure 17.2 shows an inverter that is used to convert the 12VDC into AC but on a much smaller scale than a PV system. The inverter shown has a maximum output of 500W, but it can run at that maximum only for about 20 seconds before its internal system shuts it down. That means this particular inverter can run my laptop but not my gaming PC. It also has USB ports to charge my phone or other small devices. Bear in mind, this is a small, inexpensive inverter, which is not designed to run a whole house but works great to run my laptop from my car battery if I'm camping in the middle of nowhere. It has connections for both a car's internal 12V socket (formerly known as the cigarette lighter) and cables to connect to a car battery. It's able to go from 12VDC to 110VAC by first inverting the signal from DC to AC and then using a transformer like the one we learned about in Chapter 16 to boost the 12 volts to 110 volts. For anyone who doesn't have a PV system, an inverter like this would be an appropriate tool to have in case of a long power outage, so they could charge their device to call for help or run their laptop from their car and keep up with news of the event. Bear in mind that some newer trucks and SUVs or minivans come equipped with 110VAC plugs, just like the wall outlets in your home, but it's not a common feature. . .yet.

Figure 17.2: A small inverter

For computer geeks, UPS is more than a shipping company. A UPS is an uninterruptable power supply and is used to provide conditioned AC power output from a backup battery (until a backup generator comes online) when 100% uptime is required, such as in a hospital, or merely to keep a computer and its monitor running long enough to shut down a system without loss of data. The inverter in Figure 17.2 relies on an external battery, but a UPS has an internal battery whose charge is constantly maintained by the device while the mains power is on, and it switches over to using battery power and the inverter in milliseconds when it senses that the mains power is no longer available. A UPS can range in size from about the size of the inverter in Figure 17.2 to support a single computer or device to something the size of a refrigerator or bigger for a commercial concern.

Rectifier Uses

Rectifiers are the heart of many power supplies. Consider the power supply in a desktop computer. PCs use voltages of 3.3, 5, and 12VDC. A PC's power supply takes the 110–120VAC from a wall outlet and drops it down using a step-down transformer with multiple taps or windings on the secondary side. Then it uses diodes, capacitors, and other devices to rectify the AC signal to DC and smooth out the ripples, and it uses a voltage regulator to ensure the voltages are at the proper level. Some use a buck converter, which is a DC-to-DC converter to step voltage down.

Rectifiers are not just for power supplies. Consider an OLED TV. The O in OLED is for organic, meaning that the device is made in part from carbon, the element that exists in every organic (living) thing. In an extremely simplified explanation, an OLED has a grid of transistors, each of which determines whether a pixel on an OLED is on or off. We all know that transistors run on DC; therefore, an OLED needs a rectifier to convert the household AC that comes from our wall outlets into DC that the OLED's transistors can use.

Many of the devices in our homes (i.e., anything with a circuit board) most likely contain a rectifier, because while our power is received via AC as it's easier to transmit it that way, many devices, especially electronics, require DC to run.

Construction of Inverters

Inverters rely on oscillators. As mentioned in Chapter 16, an oscillator can be created in several ways. An inverter may use two transistors and a few resistors like the multivibrator circuit in Chapter 10, "Capacitors," it could use a relay circuit like the relay oscillator in Chapter 11, "The Magic of Magnetism," or it could be like the 555 timer circuit in Chapter 16. The oscillator is then coupled with a transformer to boost the voltage. See Figure 17.3.

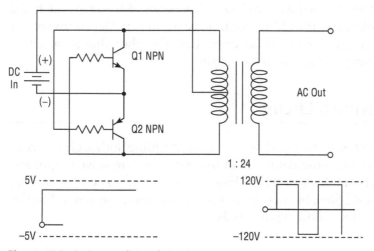

Figure 17.3: An inverted signal circuit

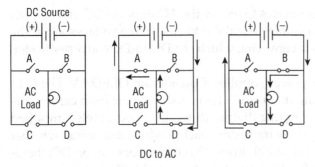

Figure 17.4: Inverting current

Regardless of the method, the basic principle driving an inverter is the same. Figure 17.4 shows the basic premise of an inverter, which is explained in the following paragraph.

Inverting DC to AC requires no fewer than four switches. Power always flows from the negative terminal to the positive terminal, but when switches A and D are closed, it flows through the load in one direction. When switches B and C are closed, the current flows through the load in the opposite direction. What changes from one inverter circuit to the next is the device used to control the alternate opening and closing of the switches. The switches alone produce an alternating current in the form of a square wave. However, a much better sine wave, like one that would come from a generator, can be produced using pulse width modulation. Inductors, resistors, and capacitors can be added to the circuit to smooth out the signal and this is called *passive filtering*. There is also *active filtering*, but that requires the components to have their own external source of power.

Often, we need an inverted wave to have a frequency of 60 Hz to match U.S. standard household current, or 50 Hz for European current. Depending on the oscillator circuit type, the frequency can be controlled by the value of capacitors, resistors, and possibly inductors on the circuit.

Try This: Filtering a Circuit

In this activity, we'll use the PWM feature of an Arduino, although we could also use the output of an oscillator circuit to create a square wave output and then filter that output into something closer to a sine wave. The filter we'll be using is called an *RC filter*, because it uses a resistor and a capacitor. For this project, you'll need the following materials:

■ Oscilloscope

■ Breadboard

■ Arduino Uno and USB cable

- Arduino IDE
- (4) 100-ohm resistors
- (4) 2.2 uF capacitors
- 47 uF capacitor
- 100 uF capacitor
- Jumper wires
- LED

Setting Up the Circuit

Use Figure 17.5 to verify your circuit configuration. The breadboard points referenced are those of the demo circuit. Your breadboard might be different, so adjust the reference points as necessary.

Figure 17.5: An Arduino PWM circuit

1. Place the negative lead of an LED in 7f and the positive lead in 3f.
2. Jumper from 7a to the ground rail.
3. Place a 100-ohm resistor between 3d and 1d.
4. Jumper from 1a to PWM pin 10 on the Arduino Uno, and from the ground rail to the GND pin on the Arduino Uno.

5. Open the IDE and create a new file. Call it PWM Filter Circuit. Write the following code in the `void loop()` section:

```
analogWrite(10, 128);
  delay(500);
```

6. Save the sketch and then compile and upload it to the Arduino Uno. The LED should appear to stay lit steadily. This code sets the duty cycle to 50% and will output a nice square wave. The frequency on PWM pin 10 is approximately 490 Hz.

7. Connect the oscilloscope's black lead to the negative LED lead, and the oscilloscope's red lead to the positive side of the LED. Adjust the oscilloscope's *volts per division* and *seconds per division* until a clean square wave appears. See Figure 17.6.

8. Disconnect the jumper from 1a and unplug the Arduino.

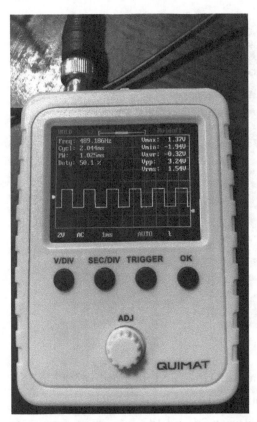

Figure 17.6: A square wave

Filtering the Circuit

The filter we will create is called an RC filter, and it's a low band-pass filter, which means it will filter out all frequencies above a certain frequency. We must calculate the cutoff frequency to ensure that with the filter, our frequency will be allowed to pass. The formula for the cutoff frequency is as follows:

$$F_{cutoff} = \frac{1}{2\pi RC}$$

We know that the cutoff frequency needs to be above the 490 Hz supplied by the PWM pin 10 on the Arduino, so begin by assuming a cutoff frequency of 1,000. Remember that resistance (R) must be expressed in ohms, not kilo ohms, and capacitance (C) must be expressed in farads, not microfarads. We don't want the resistance to be so high that the LED won't work, so we'll assume we're installing a resistor of 100 ohms and turn the formula around.

$$C = \frac{1}{2\pi RF}$$

Replacing R with 100 and F with 1,000 works out to be 1.59 uF. Because that's not a standard size and 1,000 Hz is higher than needed, if the 1.59 uF is rounded up to 2.2 uF (a standard size), then the cutoff frequency becomes 723 Hz, which will work for our circuit.

To add the first filter to the circuit, simply place a 2.2 uF capacitor so that its positive lead is in the same terminal row as the limiting resistor and the negative lead connects to the ground rail. Figure 17.7 shows how to add the RC filter to the circuit. With the first filter, there is little change, but as second and third resistor and capacitor pairs (stages) are added, the signal becomes a nice sine wave. Figure 17.8 shows the circuit schematic and image with a four-stage filter added.

To add subsequent filter stages, move the LED to the right, and insert a resistor so that one end is in the same terminal row as the previous resistor, and the other is in a row where the LED's positive lead will be. Then place a 2.2 uF capacitor so that its positive lead is in the same terminal row as one end of the resistor, and the negative lead is in the negative power rail. Again, refer to Figure 17.8 to see the four-stage filter in place.

Figure 17.9 shows the results that were achieved on the demo circuit using various filters as noted. Note also that although a good sine wave was achieved, adding each filter resulted in a lower voltage at the load and a dimmer LED.

Figure 17.7: A single RC filter stage

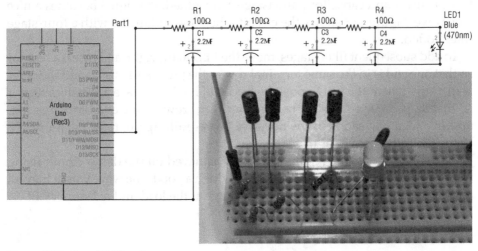

Figure 17.8: Four RC filter stages

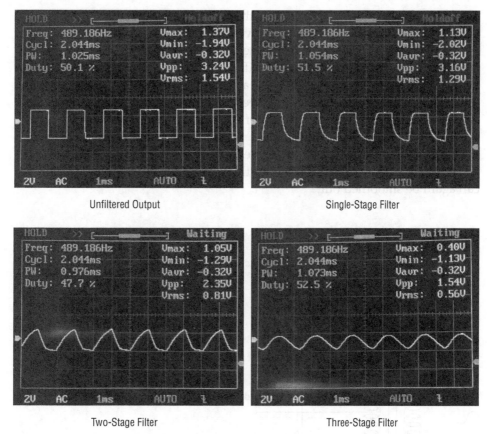

Figure 17.9: Filtered oscilloscope outputs

Construction of Rectifiers

As previously discussed, rectifiers do the exact opposite of inverters. A rectifier converts an AC wave to DC through the use of diodes. Different types of DC signals can be created from an AC wave, depending on how the diodes are arranged.

Figure 17.10 shows three types of rectifiers. The first, Figure 17.10A, provides half-wave rectification. Because there is only one diode, the current is allowed to flow in only one direction, and the opposite side of the wave is chopped off. Half-wave rectification does provide DC because the output flows in only one direction (never negative). But it is pulsating DC because the voltage pulses between its maximum and zero; it is not constant.

Figure 17.10B shows full-wave rectification using a center tap transformer. When the current flows in the first direction, it takes one path and is blocked by the opposite diode. When the AC current reverses direction, the opposite diode conducts. This results in the bottom of the AC wave going in the same direction across the load; therefore, it's a DC wave. The voltage is very uneven, but, just like an inverter circuit, it can be smoothed out using capacitors or other devices.

Figure 17.10C shows full-wave rectification using a bridge rectifier. These are fun to build but can be purchased as complete units to save the trouble of building one. Bridge rectifiers don't need a center tap. In fact, the transformer can be replaced by a wall plug. Two of the diodes conduct electricity during one half of the AC wave, and the other two diodes conduct during the second half of the AC wave. However, regardless of the direction of AC, the direction across the load is always the same, so this provides full-wave DC output that can again be smoothed out.

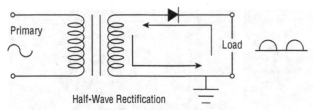

Figure 17.10A: Half-wave rectification

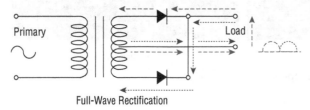

Figure 17.10B: Full-wave rectification, center tap transformer

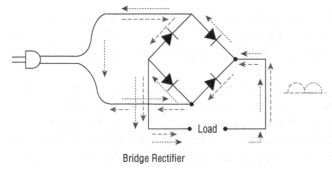

Figure 17.10C: Full-wave rectification, bridge rectifier

Single-Phase vs. Three-Phase Power

While we're talking about bumpy output waves, let's take a look at single-phase versus three-phase power. Figure 17.11 first shows a single-phase generator. As the magnetic field turns within the windings of the generator, the current flows first one direction and then the other. With just one set of windings, the output will be a sine wave. It's the AC that makes our world work, but it doesn't have a steady voltage output, which can cause problems with electrical equipment.

Three-phase power uses three windings of conductors spaced at an equal distance around the rotor, making them 120 degrees apart. As the magnetic field moves across each, they generate a sine wave, but the peaks of the three sine waves together create a more consistent voltage output than a single sine wave. Another advantage of three-phase power is that, depending on whether the windings are connected in series or parallel (called delta and wye), different voltages can be created. Industrial equipment often uses three-phase AC power.

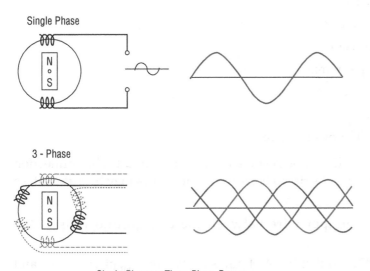

Single-Phase vs Three-Phase Power

Figure 17.11: Single-phase and three-phase power

Try This: Building a Small Variable Power Supply

Most of the projects that we've completed in this book use a 9V battery. But there may be times when a circuit calls for a 6VDC input, a 12VDC input, or something that can't be easily created by connecting batteries in series. This is where a variable power supply comes in handy, so this project walks you through the steps of creating a small variable VDC power supply using a popular voltage regulator, the LM317 in a TO-220 package. It doesn't invert or rectify the voltage, but it does allow you to change the voltage easily.

The following materials are required for this project:

- Breadboard
- LM317T
- Pair of alligator clips with conductors, red and black
- Arduino Uno, USB cable, computer with IDE installed
- 0.1 uF capacitor
- 10 uF 50V electrolytic capacitor
- 1 uF 50V electrolytic capacitor
- 1N4002 diode
- 240-ohm resistor
- 5k potentiometer
- 100k-ohm resistor
- (2) 10k-ohm resistors
- Assorted jumper wires
- Mini breadboard
- LED or buzzer
- 9VDC or 12VDC power source

Figures 17.12 and 17.13 can help you assemble this circuit. The connection points referenced are for the demo breadboard, but yours may be different, so adjust accordingly.

1. Place a black jumper wire between the two ground rails of the breadboard.

2. Insert the LM317 so that the ADJ pin is in 10a, the V_{out} pin is in 11a, and the V_{in} pin is in 12a. Refer to the pinout in 17.12 or the datasheet.

3. Insert the 0.1 uF capacitor so that one lead is in pin 12b and the other in 13b.

4. Jumper from 13a to the ground rail.

5. Place the 1N4002 diode so that the anode is in 10b, and the cathode is in 11b. You'll need to bend it and trim the leads so it stands upright between the two points.

6. Insert the 240-ohm resistor so that one lead is in 10c, and the other is in 11c standing upright like the diode.

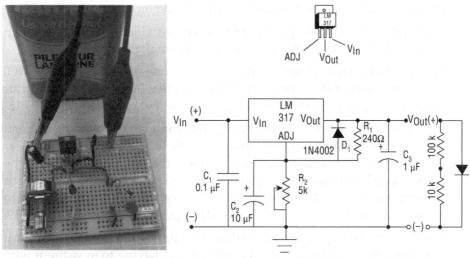

Figure 17.12: Variable power supply circuit

7. Jumper from 12d to 17d and then from 17c to the power rail.

8. Insert the 10 uF capacitor so that the positive lead is in 10d, and the negative lead is in 5d.

9. Jumper from 5a to the ground rail.

10. Insert the 1 uF capacitor so that the positive lead is in 11f, and the negative lead is in 11g, crossing over the dip.

11. Jumper from 11k to the ground rail.

12. Insert the 5k potentiometer so the dial is facing away from the other components to make turning it easier. One outside lead should be in 3i, the middle lead in 5i, and the other side lead in 7i.

13. Jumper from 7g to 7f, crossing the breadboard dip.

14. Jumper from 7e to 10e.

15. Jumper from 11e to 20e.

16. Jumper from 20f to 20g, crossing the dip.

17. Insert a 100k-ohm resistor between 20h and 22h.

18. Insert a 10k-ohm resistor between 22i and 25i.

19. Jumper from 20k to the power rail.

20. Jumper from 25k to the ground rail.

21. Jumper from 5k to the ground rail.

22. Test the potentiometer with a multimeter to verify that the two poles (the middle and one outside) connected to the circuit are at a minimum resistance, turning the potentiometer as needed.

23. Insert an LED with its positive lead in 20k, and its negative lead in 25k.

24. Cut two pieces of hookup wire approximately an inch long and then strip the entire length of each. (It's easier to strip small chunks at a time.)

25. Bend each lead in half, inserting both ends of one wire in the positive rail, and both ends of the other wire in the negative rail. These will be your connection points to attach the alligator clips. Refer back to Figure 17.12.

26. Connect a power source to the voltage regulator circuit, clipping one end of the black alligator clip and conductor to the negative rail, and the other end to the negative of the source. Connect the red alligator clip to the positive rail, and the positive end of the power source. Alternately, if your power source has lead wires, they can be plugged into the rail. Be sure to use the rail that is attached to the LM317. Again, refer back to Figure 17.12.

27. *Slowly* turn the potentiometer until the LED lights. You'll want to see what voltage is being applied to the circuit without having to measure it with a multimeter every time, so we'll go back and do the same procedure we did with the voltmeter in Chapter 5, "Dim the Lights."

28. Pull the LED out of the circuit.

29. Measure the voltage drop across both resistors together, so from the resistor lead in 20h to the resistor lead in 25i, and jot it down.

30. Measure the voltage drop across the 10k resistor and jot it down.

31. Disconnect the power from the circuit. The total voltage divided by the voltage across the 10k resistor gives the factor to use in the voltmeter.

32. Open the IDE and then open the voltmeter sketch that you saved from Chapter 5. (You did save it, right?) If you didn't save this sketch, go back to Chapter 5 and follow the instructions to create a voltmeter; then continue.

33. In the sketch, change the value in the line that reads `float const factor=11.03;` (or whatever your factor worked out to be) to the value that was calculated earlier by dividing the total voltage by the voltage drop of the second resistor. This time, the demo circuit turned out to be 12.22. Yours will be different, depending on the inherent variances in resistors and circuit components.

34. Refer now to Figure 17.13. With the power disconnected, jumper from the breadboard position 20g (between the two resistors) to Arduino pin A5.

35. Jumper from the ground rail to the GND pin on the Arduino.

36. Compile and upload the sketch to the Arduino.

37. Open the Serial Monitor. You should see values of all zero scrolling.

38. Reconnect the power to the voltage regulator circuit to test what you just did. The voltage displayed should be very close to what was measured earlier across the two resistors. Feel free to measure the voltage drop across the two resistors together now to verify what you're seeing on the screen. This voltage is not the voltage applied to the circuit; it is the output voltage that will be used to power some other circuit. Parallel branches of a circuit will have the same voltage across each branch, so the circuit being powered by your power supply will be connected in parallel with the resistors and should have the same voltage that is measured across the two resistors.

39. Disconnect the power to the voltage regulator circuit.

40. Insert a long red jumper into the power rail closest to the resistor pair.

41. Insert a long black jumper into the ground rail closest to the resistor pair. These jumpers can be left here to be power input to your other circuits. When you're ready to power a circuit, simply connect these two jumpers to the power and ground rails of the other circuit.

42. Insert the second 10k resistor into a mini breadboard with the ends of the resistor in different terminal rows.

43. Plug the opposite end of the red jumper from the voltage regulator into the same terminal row as one end of the resistor, and the black jumper into the same terminal row as the other end of the resistor.

44. Reconnect the power to the circuit.

45. As you turn the potentiometer dial, stop at various positions and measure the voltage drop across the resistor on the mini breadboard, and compare it to the voltage reported in the Serial Monitor.

If you find that regardless of what you're measuring, the voltage on the Serial Monitor is different than the voltage measured on the circuit, it may be the resistance of the conductors between one board and the other. To counteract this, adjust the value in the `float const factor=11.03;` line as needed.

This is not intended to be a high-voltage power supply, so bear that in mind when choosing the voltage for the input side of the voltage regulator. The difference in voltage between input and output is dissipated as heat, so when the voltage out is much lower than the voltage in, it will get quite hot. Consider adding a heat sink to dissipate the heat.

Figure 17.13: A variable power supply

The demo circuit was tested with a 17V input that yielded about a 16V maximum output, so the actual voltage reaching the Arduino would only be about 1/10 of that, approximately 1.6V, well within what the Arduino is capable of. And 16V with 110,000 ohms of resistance would yield a current of less than 1mA reaching the Arduino. The LM317 has a maximum voltage of 40VDC, and the original voltmeter allowed for a voltage just a bit higher than that, so my suggestion is to stay well below the 40VDC limit of the LM317.

This circuit would be even better with an LCD screen displaying the voltage instead of using the Serial Monitor. That way, an alternate source could be used to power the circuit instead of your PC, and the circuit would be more mobile. Working with an LCD display was explained in Chapter 6, "Feel the Power," so now would be a great time to practice that skill!

CHAPTER 18

Radio Waves and Tuned Circuits

"The opinion of the world does not affect me."
—Nikola Tesla

If you have not read about Nikola Tesla, now would be a great time to catch up on him. Most call him a genius ahead of his time. A few call him a madman, but he was a prolific inventor, and although he wasn't credited with it until after his death, Nikola Tesla invented the radio and was granted the patents for it. Tesla didn't care about the opinions of other people. He trusted himself, his brilliance, and his dreams no matter what his critics said, and as a result, he invented some of the greatest technological advances the world had seen, which still affect us today.

Radio Waves

If you've ever thrown a stone in a still pond, then you've seen waves. The waves look like the sine wave of alternating current, undulating up and down. We are bombarded continually with mostly invisible waves of energy, like the waves in a pond. Take a look at Figure 18.1, the electromagnetic spectrum. The electromagnetic spectrum is a range of energy waves of varying frequencies and wavelengths. (Note, frequency and wavelength are the inverse of each other.) Only a small portion of the energy we're bombarded with constantly is visible to us as color. The rest ranges from gamma rays with high frequency and energy to radio waves with low frequency and long wavelengths. While the waves on the gamma ray end of the spectrum can be deadly and easily pass through human tissue, waves on the radio end of the spectrum are relatively benign.

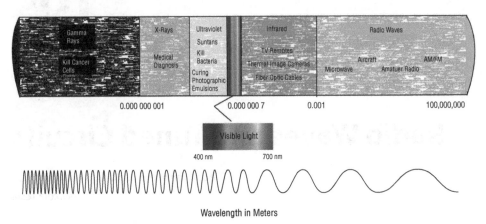

Figure 18.1: The electromagnetic spectrum

An electromagnetic wave occurs when there is a charged particle. A charged particle can be essentially pushed by a similarly charged particle, and when it moves, a magnetic field is produced. This magnetic field can also induce the charged particle to move, and this oscillation between the magnetic field and the electrostatic field creates waves of light energy, some of which are visible. The electromagnetic spectrum derives its name from the fact that the energy on it is a dance between an electrical field and a magnetic field.

We know that when a magnetic field and a conductor have relative movement between the two, a current is induced to flow in the conductor. Keeping this in mind, consider the waves of electromagnetic energy striking a conductor. Wouldn't that induce a current to flow? It does. When the conductor is a radio antenna, all that is truly needed to pick up a radio station is the antenna and a few simple parts to harness the sound. A simple radio, often called a *crystal radio*, can be constructed of an antenna, inductor (coil), capacitor, diode, and earphones. No external energy source is required, because the power for the radio is received through the electromagnetic waves striking the antenna. The sound it produces is somewhat quiet and can be greatly enhanced by an amplifier, but the amplifier isn't required for the radio to function.

Try This: Building a Radio Receiver

As previously mentioned, it doesn't take much to build a simple radio receiver. These simple radios are often called crystal radios because one of the main components, a diode, was, for a time, made from crystalline minerals. While a crystal radio may be a great learning tool and solidify the understanding that external power is not needed, it's also sometimes difficult to find the parts. The earpiece is eerily old school, and the volume will be very low.

Instead of the traditional piezoelectric earpiece that is used with a crystal radio, I prefer to add an amplifier circuit and an 8-ohm speaker. The amplifier and speaker do require the outside power of a 9V battery, but again, the radio would work without any external power; it just isn't very loud with the earpiece. The amplifier circuit uses an LM386N-1 operational amplifier (op-amp). For a refresher on op-amps, refer to Chapter 13, "Integrated Circuits and Digital Logic."

Also, note that a single integrated circuit chip can be purchased that has AM/FM radio capabilities, so the crystal radio is more for learning or the satisfaction of building it. For a bigger challenge, check out the ways to make a variable capacitor and piezoelectric earpiece. See `cliffjumpertek.com/radio` for more information on those.

For this project, you will need the following materials to make a crystal radio:

- 30 feet or more of 22 AWG insulated wire (hookup wire)
- A roll of 30 AWG magnet wire
- 5 inches of 3/4″ PVC or something with a similar diameter—I used a piece of foamboard cut and folded to yield the same diameter
- Sandpaper
- Diode—1N4148 or 1N34 germanium, but almost any diode will do
- 100k-ohm resistor
- Variable capacitor—224pf or higher
- High-impedance speaker or earpiece
- Soldering iron and solder

If you choose to build the amplifier circuit instead of using the earpiece, you'll also need the following materials:

- LM386N-1 op-amp IC
- 9V battery and connector
- Switch
- 10k-ohm potentiometer
- 10 uF capacitor
- 220 uF capacitor
- 0.047 uF capacitor
- 10-ohm resistor
- 8-ohm speaker
- Jumper wire

Figure 18.2 shows the schematic and connections for the radio and amplifier circuit. The output of the radio becomes the input for the amplifier circuit. Building a crystal radio is an imprecise science, so values may need to be tweaked a bit to pick up a radio station.

Use the inset of Figure 18.2 to assist in assembling the crystal radio.

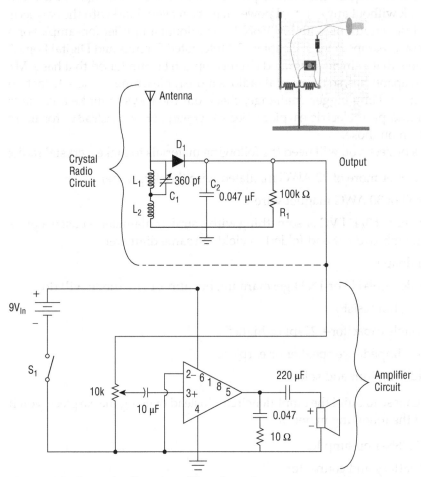

Figure 18.2: A crystal radio and amplifier schematic

1. Begin by sanding the end of the magnet wire to expose the copper. Leaving a lead of about 6 inches, tape the magnet wire to its post, and then begin wrapping the magnet wire around the pipe, wrapping each turn as close as possible to the previous turn of the wire, for approximately 30 turns to make L_1.

2. At 30 turns, twist the wire to make a loop about 0.5 inch in diameter. This loop will stick out from the wrapped wire so it is accessible. Then go back

to winding the wire, this time winding 70 turns to make L_2. Varying the number of turns in L_2 will change the frequency, so you might want to make a loop every 10 turns or so to experiment with.

3. Tape the magnet wire to the post, again leaving about a 6-inch lead, and then cut the wire.

4. Sand the loop between L_1 and L_2. This will be the connection point for the variable capacitor.

5. Sand each end of the magnet wire to expose the copper wire inside.

6. Sand each loop of the magnet wire to expose the connection points.

7. Solder one end of the variable capacitor to the connection point between L_1 and L_2.

8. Solder the other end of the variable capacitor to the opposite end of L_2 and the positive lead of the diode.

9. Solder the antenna wire to the loose end of L_1.

10. Solder another piece of wire to the connection point between L_1 and L_2. This is the ground wire.

11. If using the high-impedance earpiece, solder one wire from the earpiece to the negative lead of the diode.

12. Solder the other wire from the earpiece to the connection point between L_1 and L_2.

13. Connect the ground wire to the physical ground by literally sticking it into the dirt, or connect it to a metal pipe going into the ground.

14. String the antenna as high as possible, but if it is touching a tree or other object, be sure to place some nonconductive material between the antenna and the object. A piece of folded cardboard might work well.

Now comes the fun part. Slowly adjust the variable capacitor until you pick up sound. It will be faint, but it should be there. The closer the circuit is to a radio station, the louder the signal will be. You might need to adjust the capacitor, the antenna position, or the connection point on L_2 to receive a signal.

If you prefer to add an amplifier to the circuit and use a speaker, follow these directions, using Figure 18.3 as a guide. Your breadboard points might be different, so adjust as needed.

1. Unsolder, or clip off, the earpiece if previously attached to the circuit.

2. Place an LM386 with pin 1 in 10e and pin 8 in 10f so that the chip straddles the breadboard dip.

3. Place a switch with the center pin in 3h, and place an outside pin in each of 2h and 4h.

4. Jumper from the center pin at 3f to the ground rail.

5. Jumper the IC pin 2 (11a) and pin 4 (13a) to the ground rail.

6. Jumper from the IC pin 6 (12j) to the power rail.

7. Place a 220 uF capacitor with its positive lead in 13j and its negative lead in 14j.

8. Jumper from 14h to 28h.

9. Connect the positive lead of the 8-ohm speaker to the capacitor by inserting the lead in 28f.

10. Connect the negative lead of the 8-ohm speaker to the ground rail.

11. Jumper from the IC pin 5 (13h) to 15h.

12. Place a 0.047 uF capacitor between pins 15g and 16g.

13. Connect a 10-ohm resistor between 16h and 22h.

14. Connect 22j to the ground rail.

15. On the other side of the IC, place the 10k potentiometer with its center pin in 22b, and place an outside pin in each of 20b and 24b.

16. Jumper from the IC pin 3 (12d) to 16d.

17. Place a 10 uF capacitor with its positive lead in 16b and its negative lead in 17b.

18. Jumper from the capacitor at 17d to the potentiometer's middle lead at 22d.

19. Jumper one of the potentiometer's outside leads to the ground rail.

20. The other outside potentiometer lead will connect to the output of the crystal radio. You should connect it to the negative lead of the crystal radio's diode, and place a jumper from the ground rail of the amplifier to the ground connection between the two coils.

21. Jumper from the positive rail on one side of the breadboard to the positive rail on the other.

22. Jumper from the negative rail on one side of the breadboard to the negative rail on the other.

23. Ensure that the switch is in the off position; then connect the 9V battery, inserting the positive lead in a positive rail and connecting the negative lead to the outside of the switch at position 2g.

This completes the amplifier part of the circuit. The default gain on the LM386 is 20, but connecting pin 1 to pin 8 using a resistor and capacitor can enhance the gain up to a gain of 200.

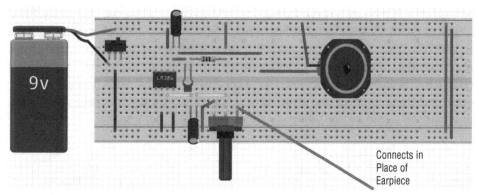

Figure 18.3: Amplifier circuit complete

The crystal radio is often a beginner project and obviously is a simplified version of a radio to assist with teaching radio concepts. An entire book could be written about the different aspects of radio waves and how to use them. However, a bird's-eye view of radio systems follows; then we'll connect an Arduino FM radio and create an FM transmitter circuit to connect to it.

Making Waves

When a human yells, how far does the sound travel? Unless the sound is echoing off canyon walls, it will travel just under 600 feet if the yell is very loud. Sound waves by themselves are mechanical waves from the pressure of moving air, like the ripples in a pond, rather than electromagnetic waves. So, how is it that radio waves are on the electromagnetic spectrum? To send sound using the electromagnetic spectrum, an electrical oscillator is needed. The wave resulting from the oscillator is called a *carrier wave*, because it *carries* the sound waves in one of two forms: either *amplitude modulation (AM)* or *frequency modulation (FM)*.

Transmitting Radio Waves

Figure 18.4 shows a simplified block diagram for a radio transmitter. It begins with a source of sound, which might be someone speaking or playing music. The sound waves are intercepted by a transducer, such as a microphone, and converted to an electrical voltage.

Meanwhile, an oscillator creates a sine wave of a specific frequency within the radio frequency portion of the electromagnetic spectrum. This frequency is the frequency of the radio station; for example, "tune to 1020 on your AM dial" would mean that the radio station being talked about transmits at a frequency of 1020kHz. If the radio waves are on an AM frequency, the oscillations will be in the range of 535kHz to 1605kHz, while FM stations operate between 88MHz to 108MHz. These frequencies, and the power that can be used to transmit them, are controlled in the United States by the Federal Communications Commission (FCC). Radio stations are licensed for a certain frequency band in a geographic area, and these licenses are granted and controlled by the federal government to ensure that one radio station doesn't overlap another's signal.

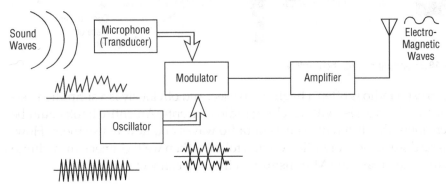

Figure 18.4: A simplified radio transmitter

At the modulator of our transmitter, the sound waves and carrier waves are combined. Along the way, several amplifiers will boost the voltage of the sound, which will be sent out over an antenna as a modulated wave. The more power that is used to send the signal, the farther the waves will travel. But don't worry, the radio frequency devices we create here are not powerful enough to get you into trouble with the FCC.

Receiving Radio Waves

When radio waves are received by an antenna, the process is essentially reversed. On the receiving end will be a bandpass filter, which uses a combination of an inductor and capacitor to determine the frequencies it will accept and pass to the next part of the circuit. Any other radio frequencies will be ignored. Figure 18.5 is a simplified block diagram of the receiving end of a radio signal.

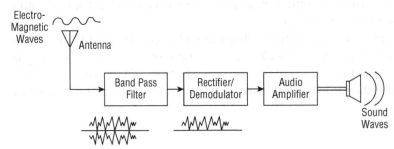

Figure 18.5: A simplified radio receiver

After the proper frequency signal is accepted, the signal is rectified by a diode so that only half of the original signal remains. The signal is also demodulated, meaning that the carrier wave is filtered out and what remains is the sound signal, which is amplified and sent through another transducer like a speaker or headset.

AM vs. FM

As mentioned previously, AM radio waves change the amplitude of the carrier wave, while FM radio waves change the frequency. Figure 18.6 depicts what the waves might look like if they were visible.

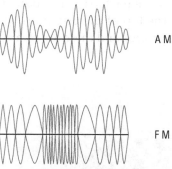

Figure 18.6: AM and FM waves

AM stations are more susceptible to interference from other electromagnetic waves. However, FM stations are largely unaffected by stray *electromagnetic interference (EMI)*, because while extra energy will change the amplitude of a radio wave, it won't affect the frequency. EMI can be caused by microwaves, cell phones, trees touching power lines, lowly light switches, and even organic sources of energy, such as sunlight and lightning.

Other environmental factors can cause a radio signal to travel farther, instead of causing interference. Radio waves will bounce across water much like a stone skipping across a pond. Radio waves will also bounce off the atmosphere and back to Earth, which is how we are able to send radio waves from one continent to another.

Regardless of whether the amplitude or the frequency is modified by adding the carrier wave, when the carrier wave is removed again and rectified back to only one-half, what's left is the sound wave.

Try This: Building an Arduino FM Radio

Arduino shields were mentioned in Chapter 4, "Introduction to the Arduino Uno," but this is the first project in which we'll be using one. As a refresher, many shields are available for Arduino. A shield is a circuit that has been completed by someone else, or a company, and often includes a library. Shields can save time, money, and frustration, but they do eliminate some of the joy of figuring out the circuit on your own.

The Shield

The shield that we'll be using in this project was purchased from an online merchant for less than $10 U.S. It is an FM radio with an antenna promising high sensitivity and stability with low noise. It operates from 76MHz to 108MHz, which includes the FM radio band.

The manufacturer of this shield is Ximimark, and the shield is based on a TEA5767 FM radio chip (the square chip when looking at the top of the module). See Figure 18.7. The other chip and circuitry on the top of the board are an amplifier circuit like the one shown in the previous project, and these are clearly surface mount components.

Figure 18.7: The TEA5767 FM radio shield

On the other side of the chip, the two 3.5mm sockets are labeled for sound output and an antenna.

The four pins that stick out the opposite side of the module are for communication using a standard called I2C (pronounced "eye squared see") and also a +5V power and ground connection. The pin labeled SDA is the data pin that actually sends information between the Arduino board and the shield. The other pin labeled SLC is a clock pin that keeps the Arduino board and the shield working in sync. You'll see these same pins on other devices that use the I2C communication standard, such as organic LED (OLED) displays.

The Libraries

As we learned in Chapter 4, libraries add extra functionality to an Arduino sketch and save the user a great deal of time, because someone else has done the necessary programming in C or C++. It's fun to create new things, but why would we spend time to solve a problem that someone else already has? We wouldn't, and we aren't going to. (We're also not learning C++ programming here, because that would be entirely another book.)

The shield manufacturer didn't supply a library, so in this project we're going to take advantage of a couple of open source libraries: one called *wire* and another called *TEA5767Radio*. Wire is used to communicate via the I2C standard and is specific to the hardware board being used, so it should already be available in your integrated development environment (IDE). The other library, written by Simon Monk, is used to access the TEA5767 chip. Both of these libraries are available for free in the Arduino IDE. More information about the wire library is available in the reference section of the Arduino website.

Refer to Figures 18.8 and 18.9 to assist in performing the following steps, which will install the necessary libraries into your system:

1. Open the Arduino IDE on your computer.

2. Select File ➪ New to open a new sketch.

3. In the new sketch, select Sketch ➪ Include Library ➪ Manage Libraries (see Figure 18.8). A search box will open (see Figure 18.9).

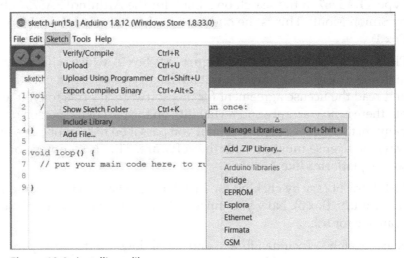

Figure 18.8: Installing a library

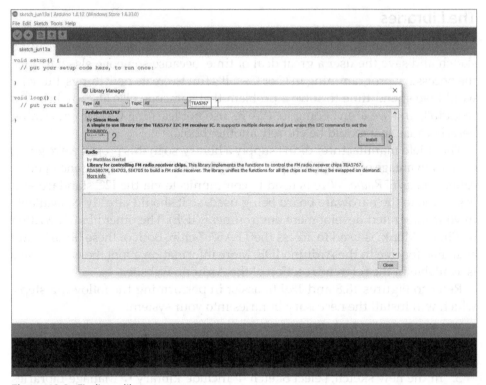

Figure 18.9: Finding a library

4. Type **TEA5767** in the search box. Look for the ArduinoTEA5767 library by Simon Monk. This is the one used in the example here (Figure 18.9, Box 1).

5. Please click More Information (Figure 18.9, Box 2), which will take you to `https://github.com/simonmonk/arduino_TEA5767,`. Then click License and read the license agreement for this software. It's free for you to use, but there is no warranty and no liability on the part of the person or company that wrote the software, and if you pass it on to someone else in any way, the license must go with the software. This is common for open source platforms like Arduino.

6. Install the library by clicking the Install button back in the Arduino IDE (Figure 18.9, Box 3). Now you must add the library to the sketch so it is available for use.

7. Select Sketch ⇨ Include Library, and scroll down to the library that was just installed. It should be under Contributed Libraries. Click its name to include it in your sketch.

8. To add the wire library to the sketch, type the following line at the top of the sketch: `#include <wire.h>`.

Verifying the Radio Works

For this project, you'll need the following materials:

- Arduino Uno and USB connector
- Breadboard
- PC with Arduino IDE installed
- Headphones with standard 3.5mm connector
- Jumper wires
- 10k potentiometer

We'll start off using the example code to verify that the radio module works. This sample code is provided with the TEA5767 radio library.

> **NOTE**   If you would like to display line numbers, in the IDE, select File ≫ Preferences and choose Display Line Numbers, as shown in Figure 18.10. As your sketches grow in complexity, the line numbers may prove helpful.

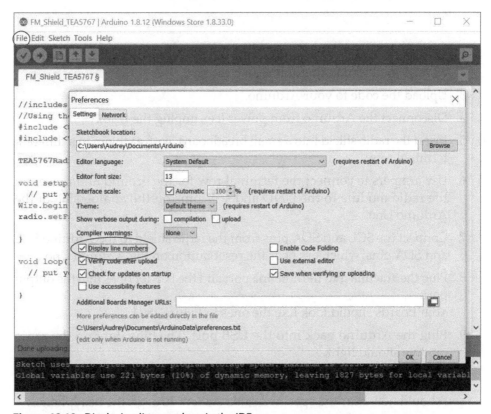

Figure 18.10: Displaying line numbers in the IDE

Use Figure 18.11 to help make the connections. First, let's look at the code.

1. Plug the Arduino Uno into the USB port and enter the following code into the IDE:

```
//includes the library to run the communication and the TEA5767
shield

#include <Wire.h>
#include <TEA5767Radio.h>

TEA5767Radio radio=TEA5767Radio();

void setup() {
  // put your setup code here, to run once:
Wire.begin();
radio.setFrequency(102.7);

}

void loop() {
  // put your main code here, to run repeatedly:

}
```

2. On line 11, change the number in `radio.setFrequency(102.7);` to the number of a local station that comes in clearly.

3. Upload the code to your Arduino.

4. Disconnect the Arduino from power (i.e., unplug the USB).

5. Insert the radio shield into a small breadboard, with each pin in a different terminal row.

6. Use jumpers to connect the terminal row for the ground and power on the radio module to the GND and +5V pins on the analog side of the Arduino Uno.

7. Connect the SCL and SDA pins from the radio shield to the Arduino SCL and SDA pins, which are near the reset button on the Arduino Uno.

8. Plug the antenna into the antenna port and the headphones into the output port.
   Your boards should look like the ones in Figure 18.11.

9. Plug the Arduino back into the USB port. The blue light on the radio module should light.

10. Either press the reset button or upload the code again to the Arduino.

You should now hear the specified radio station through the headphones.

Figure 18.11: Radio module connections

NOTE   This activity uses the Arduino Uno to minimize the number of boards needed to complete all of the activities in this book. However, a different Arduino board could be used as well. The Mega2560, Leonardo, and Due all have SDA and SCL pins.

While this is a functioning radio, it's not convenient to change the code this way each time, so this next section explains how to add a potentiometer for station selection.

Adding Station Tuning

To enable tuning to different radio stations, add a potentiometer to the circuit by performing the following steps. Remember, your breadboard may not match the demo board, so adjust the connection points as necessary. Use Figure 18.12 to aid in construction.

1. Verify that the Arduino is unplugged from the USB port.
2. Insert the radio shield into the breadboard with the +5V pin in 10f and the GND pin in 13f.
3. Jumper from 10a to the power rail.
4. Jumper from 13a to the ground rail.

Figure 18.12: Potentiometer station selection

5. Insert the potentiometer with the center pin in 28f and the outside pins in 26f and 30f. Ensure that the potentiometer is firmly seated in the breadboard.

6. Jumper from one outside pin to the power rail.

7. Jumper from the other outside pin to the ground rail.

8. Jumper from the 28c, the potentiometer's center pin, to the A3 pin on the Arduino board.

9. Jumper from 11c to the SDA pin on the Arduino.

10. Jumper from 12c to the SCL pin on the Arduino.

11. If your breadboard is split in the power rail between the sections where the radio module and the potentiometer are, jumper the two together.

12. Connect from the breadboard's ground rail to GND on the Arduino.

13. Connect from the breadboard's power rail to +5V on the Arduino.

14. Plug the Arduino into the PCs USB port.

The physical connections are complete! The potentiometer is in parallel to the radio, because both will use the same 5V connection. That means that the potentiometer could vary the output theoretically from 0V to 5V, and the value read on the input pin A3 could vary between 0 and 1,023, but as we know, components are not perfect, so the programming will make compensating adjustments.

You can begin with the previous sketch, but you will want to save it with a different name before making changes (in case you need to revert to the old one). The code will be shown in its entirety. Change your sketch to match what is shown here.

Most of the `Serial.println` lines are there for troubleshooting and can be removed once the project is working properly (although you may want to leave the last two).

```
//includes the library to run the communication and the TEA5767 shield
// and the radio module
#include <Wire.h>
#include <TEA5767Radio.h>

TEA5767Radio radio=TEA5767Radio();

int input=A3; //sets the potentiometer center pin to read from A3
int convert=0; //variable to store temporary value
float freq=0;   //creates the variable to store the chosen frequency
int min=88;  //sets the minimum value for frequency
int max=108;  //sets the maximum value for frequency


void setup() {
  // put your setup code here, to run once:
Wire.begin();
radio.setFrequency(102.7); //change this to the station you listen to
most
Serial.begin(9600);  //turns on serial monitor

}

void loop() {
  // put your main code here, to run repeatedly:

input = analogRead(A3); // reads the input on A3
Serial.println(input);

if (input<100){
  Serial.println("Below Minimum");
  }
else if (input>1000) {
  Serial.println("Maximum Exceeded");
}

convert =(input * 10)/45 + 880; //makes the frequency an integer
                                //offsetting for the minimum frequency
Serial.println(convert);
```

```
if (convert > (float)freq*10 + 1) {
   freq = (float)convert / 10;
}

else if (convert < (float)freq*10 - 1) {
   freq = (float)convert /10;
}

radio.setFrequency(freq);
Serial.println("Frequency is");
Serial.println(freq);

   delay(100);


}
```

Remember to save the code and then verify and upload it to the Arduino. Turning the potentiometer from one side to the other should now change the frequency that is selected on the radio shield.

Near the top of the code, pin A3 is set as input because it will receive a voltage from the potentiometer.

The freq variable must be set as a float because it needs to be able to address the decimal that occurs in many station frequencies, such as 104.7. The other numeric values are set to integer because the calculations are performed much faster with integers than floating points.

The min and max variables serve no useful purpose. The initial program plan was to use them in the calculations, but because they weren't used, they were left there simply as a reference for the FM frequency range.

Because no component is perfect, the code is using the input range on pin A3 only between the numbers of 100–1,000 (instead of 0–1,023) to compensate for any imperfection in the potentiometer that might leave some resistance when there should be none. The difference between 88 and 108 is 20, so the 900 range from 100–1,000 divided by 20 means each change in frequency is equal to 45 on the A3 pin.

On the convert =(input * 10)/45 + 880; line, the input is multiplied by 10 before being divided so that the number representing the radio station will be an integer, and then 880 is added to compensate for the starting frequency of 88MHz rather than 0.

Subsequent lines divide the convert variable by 10, so it will equal the frequency.

The following if/else commands won't change the input frequency unless it has varied by more than 0.1MHz, because even without turning the rotary on the potentiometer, there may be fluctuations on the input. If stations drift in and out, this value could be increased even more to make the frequency more stable.

Finally, the frequency is set to the calculated value and printed out to the serial monitor.

This project does have room for improvement. First, an OLED, or LED, like the one used in previous projects would be a great addition, because that would let the user see the frequency without being tied to a computer. The TEA5767 has many additional features right on the chip, such as scanning for the next station, a mute button, and volume, but accessing those features is a bit advanced. Take a look at the datasheet for the TEA5767 to get some ideas, then refer to Chapter 6, "Feel the Power," to configure the LCD. Once this project is working as desired, you might consider soldering it onto a perf board and making or finding a case for it.

With the addition of a wireless FM microphone circuit (they're easy to build) tuned to the same frequency, this project could be used as a baby monitor or other remote listening device.

Tuned Circuits

If you would like to know more about how radio frequencies are calculated, stay tuned! Tuned circuits are what enable a radio to select the desired frequency, and for those of you who love math, there are formulas for tuned circuits. Tuned circuits either accept or reject a *resonant frequency*.

If you've ever tried to push someone on a swing, then you have felt resonance. If you try to push them too soon, it's almost impossible, and if too late, then you may fall face first into the dirt. Yet, at just the right moment, pushing the swing is extremely easy, and the rider will go higher and faster. That's the beauty of resonance. Resonance is also what's at play when an opera singer is able to break a wine glass with just their voice. When the vibration of the sound is at the resonant frequency of the glass, the glass breaks. Massachusetts Institute of Technology (MIT) has a great video of this online at www.youtube .com/watch?v=CdUoFIZSuX0. Rumor has it that Nikola Tesla built an "earthquake machine" that could match the resonance of a building and reduce it to rubble.

Figure 18.13 shows two configurations for electrical resonance. A tuned circuit can be either an acceptor circuit or a rejector circuit, based on whether it has a capacitor and inductor in series with or parallel to each other. If the components are in parallel, the resonant frequency will be rejected, but in series the resonant frequency will be passed on to the next part of the circuit. A parallel tuned circuit is also called a *tank circuit*.

A resonant circuit is an AC, RCL circuit. This means it is an alternating current circuit that contains a resistor, capacitor, and inductor. As previously discussed, an inductor can be as simple as a coil of wire. It's important to note that whenever there is a frequency, the circuit is an AC circuit, and because most of this book is about DC electronics, there are some new terms here. I'll attempt to explain them in the simplest terms.

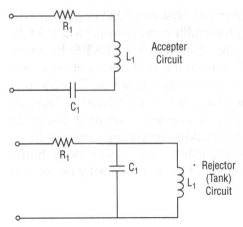

Figure 18.13: Tuned circuits

Reactance is the opposition to current from an inductor or capacitor in an AC circuit. The symbol for reactance is X, and the symbol for inductance is L, so inductive reactance is X_L, and capacitive reactance is X_C. The resonant frequency (f_o) is the frequency at which $X_L = X_C$. In this condition, if the components were perfect and no power was lost through the conductors, energy could bounce back and forth between an inductor and a capacitor indefinitely. The inductor would build its magnetic field until the current from the capacitor stopped flowing; then the inductor would discharge the electricity back to the circuit until the capacitor reached full voltage. At that time, the capacitor would return the power it was holding in its electrostatic field back to the circuit, where the inductor would take that energy and again begin building a magnetic field.

If you need to figure out the resonant frequency for a particular circuit, here are the formulas, where f is frequency, L is inductance and C is capacitance:

$$X_L = 2\pi fL$$

$$X_C = \frac{1}{2\pi fC}$$

$$f_o = \frac{0.159}{\sqrt[2]{LC}}$$

Typically with radio operations, an inductor is chosen, and a variable capacitor is used so that the resonant frequency can be changed to the desired value.

This chapter has gone from understanding the electromagnetic spectrum and the energy contained in it to radio waves and how they are controlled and made to traverse the world. After that, Arduino shields were introduced as well as using the I2C communication standard and the math behind tuning in a radio station.

In Chapter 19, "Connecting Your Circuits to the Cloud," you'll learn how to use other radio waves (Wi-Fi) to communicate with your devices via the Internet from anywhere in the world, so stay tuned!

Part

IV

Putting the I in IoT

In This Part

Connecting Your Circuits to the Cloud

"O! it is pleasant, with a heart at ease,
Just after sunset, or by moonlight skies,
To make the shifting clouds be what you please..."

—Samuel Taylor Coleridge

It's amazing what "the cloud" has come to mean. It touches our lives every day. People stay in touch with it and spend minutes finding what once would take days, and they can even remotely control their devices with it. In this chapter you will learn to make the cloud be what you please, connecting you to your device no matter where you are.

The Arduino IoT Cloud

The folks at Arduino have made it easier than ever to remotely control your devices. They have created the Arduino Create IoT Cloud, which enables you to connect devices to the Internet or to each other quickly, easily, and securely. The platform provides a method for users to configure boards and write code online, too. At the time of this writing, it is still in the beta stage, so be aware that links and such may change.

To connect to the Arduino Create IoT Cloud, you will need one of the following boards:

- MKR1000
- MKR WiFi 1010
- MKR GSM 1400

- NANO 33 IoT

- MKR NB 1500

- MKR WAN 1300

- MKR WAN 1310

The board used for the demo is an MKR WiFi 1010. Although one of the proper boards must be purchased, the Arduino Create IoT Cloud is free, with limits. The free account allows the user 100 sketches and 100MB of storage, while connecting one *thing* that can have five *properties* controlled and accessed through multiple *widgets*, and the cloud will store data for one day. (More about things, properties, and widgets in a bit.) There is also a pay-per-month plan for a nominal fee, which, of course, increases the limits of what can be connected. Finally, Arduino lists a professional plan but no prices. For that, a call to the folks at Arduino is required.

The Arduino cloud is accessed at create.arduino.cc. There you will find links called Getting Started, Arduino Web Editor, Arduino Project Hub, Device Manager, Digital Store, and Arduino IoT Cloud. This website is known as Arduino Create.

The first thing you need to do to use Arduino Create is to create an account.

Go to create.arduino.cc. The screen in Figure 19.1 should appear. Clicking either Sign In in the top-right corner or Arduino IoT Cloud in the bottom-right corner of the six blocks will bring you to a login screen where you can create an account. As with any online service, be sure to read the privacy policy and terms of service before you create your account. (Links are provided on that page.) As with many online services, they'll send you an email and ask you to click the link to activate your account before you continue.

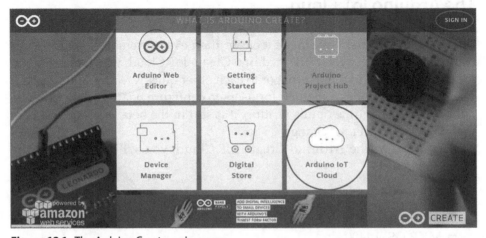

Figure 19.1: The Arduino Create main screen

If you're using a laptop or desktop PC to access the web page, it's recommended that while on the account screen you configure two-step verification under Security on your profile page at `id.arduino.cc`. Two-step verification is also called *multifactor authentication*, and it provides an extra layer of security for your account. As an IT professional, I recommend that everyone use multifactor authentication for every site that offers it. To configure two-step verification, you'll need an authenticator app on your cell phone. Look for Google Authenticator or Microsoft Authenticator from wherever you download your apps. Once installed, the authenticator application can be used for multiple websites. Install the app on your phone and then go back to the website and under Security (you have to scroll down to it) click Activate, which appears next to *2-Step Verification*, to activate multifactor authentication. Scan the QR code with your phone and follow the on-screen prompts. Just remember that from now on, you'll need your phone handy to access your account from a new device.

TIP If you're having trouble finding 2-Step Verification, navigate to `create` `.arduino.cc` and click your icon in the top-right corner. Once there, click Profile to bring you to the screen where 2-Step Verification can be configured.

You'll also need to install the Arduino Create Plugin with these steps:

1. Navigate back to `create.arduino.cc`.
2. Click Getting Started in the middle-top row.
3. Scroll to the bottom of the page.
4. Click Arduino Create Plugin to download it.
5. Click the downloaded file and install it.

Once your account is set up, it's time to add devices, things, properties, and widgets.

Try This: Setting Up Your Device

Perform the following steps to set up your device.
For this activity, you'll need the following:

- Arduino MKR WiFi 1010 or similar board
- USB cable to connect the board to the PC
- PC with Internet access
- Network name and password

1. Go to create.arduino.cc and sign in with the account you created. (If you haven't created an account yet, follow the earlier instructions for how to do so.) If you don't see Sign In in the upper-right corner, you're already signed in. The next step is to set up an Arduino board to work with Arduino Create.

2. Click the Getting Started icon in the middle-top row. Scroll down to the type of board you have and click it. This will bring up a screen to install your *device*, which is what Arduino calls your Arduino board. At this time, the free version of the Arduino IoT Cloud will allow for five devices to be connected. Ensure that you have removed the board from the conductive foam that it comes packaged in, because it's conductive and the board will short out if you apply power while it's still in the foam. Connect your device to your PC via the USB cable.

3. To connect an Arduino board to the cloud, you need to have the service set identifier (SSID; i.e., network name) and password of the router that will be used to connect to it. If your Internet service provider (ISP) is providing your router, then often the SSID and password will be printed somewhere on the router. If you're using your own router, then simply use the same network ID and password that you used when connecting your PC or phone to your Wi-Fi. Click Next on the screen and follow the on-screen prompts to configure your device. You'll name your board and then configure the crypto chip for security. The program will upload a sketch that will enable you to turn the LED on your board off and on, by clicking an icon on a screen later.

4. Click the Secret tab or Enter Wi-Fi Data button to enter the SSID (SECRET_WIFI_NAME) and password and then click the Upload button.

5. When the board is set up, a screen will come up where the LED on the board can be turned on and off, but the screen isn't particularly intuitive. Click the image of the LED to toggle the LED on and off.

NOTE   Troubleshooting: If you have problems convincing the Arduino to connect to Wi-Fi, try pressing the button on the Arduino twice and then connecting again. If that doesn't work, try cycling your router and your PC off and on again and then connecting the Arduino to Wi-Fi again. First, ensure that your password and SSID are entered exactly as on your router. If that doesn't work, you might need to change a setting on your firewall. If all else fails, search the Arduino forum at forum.arduino.cc for similar problems and try the solutions there.

6. From the flashing LED screen, the program will take you to the Device Manager. Here, you can add more devices if you want. Click the waffle in the top-left corner to get back to the main screen shown in Figure 19.1. (The waffle, similar to ones found on many websites, is a group of nine dots arranged in a square on the upper left of the screen.) That's it! You just set up your device!

Try This: Using Things, Properties, and Widgets

Now that you have your device set up, it's time to do something with it. In this project, we'll use a slider and a stepper so either one can control the brightness of an LED using PWM, which you learned about in Chapter 14, "Pulse Width Modulation," but doing it through the Internet.

For this activity, you'll need the following:

- The board set up in the previous example with its USB cord
- Breadboard
- Blue LED
- 220-ohm resistor
- Jumper wires
- PC with Internet access

First, the circuit must be configured. See Figure 19.2 and follow these steps:

1. Insert the LED so that the positive lead is in 3e and the negative lead in 5e.
2. Insert the 220-ohm resistor between 5a and the ground rail.
3. Jumper from 3a to pin A3 on the Arduino. (This is a PWM pin.)
4. On the opposite side of the Arduino, connect the GND pin (ground) to the ground rail on the breadboard.
5. Connect the VCC pin on the Arduino to the power rail on the breadboard.
6. Verify your connections by comparing to Figure 19.2. The LED will be off.

The next steps will configure your thing. From the main page in Figure 19.1, click Arduino IoT Cloud in the bottom-right corner. You should see the screen shown in Figure 19.3.

Figure 19.2: A blue LED dimmer circuit

"Things" should be highlighted on the menu bar at the top. If not, click it now. Things are shown on the left and can be any component that you want to control or monitor. Properties are characteristics or states of being of the thing in question and will be different, depending on what thing you're looking at. Widgets are used to control or read from things. (Once you have created a thing, your thing will show on this screen instead.)

1. Click Add New Thing.
2. In the Enter A Name For Your Thing box, name your thing. The demo thing is called Blue_Light.
3. Click the down arrow in the next box for a list of your devices, and choose the device that was set up in the previous project.
4. Click Create.
5. Click Add Property on the right side of the screen.
6. In the Name box, type **Brightness**. Notice that it creates a variable name called `brightness`.

7. In the Select Property Type box, click the down arrow and choose Int. Leave Min Value set at 0, and set Max Value to 255. Verify that Permission is set to Read & Write, then change Update to Regularly, and in the Every box enter **0.5**, which will check for the value of brightness every half-second while the program is running.

8. Click Add Property. Your screen should now look like Figure 19.4.

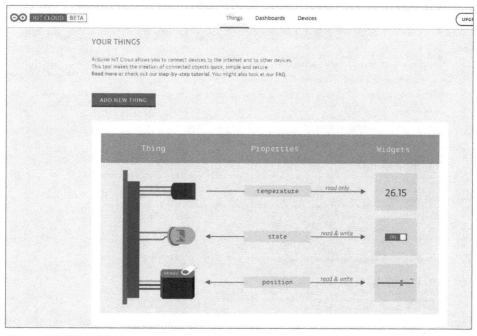

Figure 19.3: Your Things screen

Figure 19.4: The properties of Blue_Light

Next, we'll examine the code for the project. Click the Edit Sketch button. This brings you to the Editor, which will have automatically created code for you based on what was selected on the previous screens. See Figure 19.5. Ensure that the far-left tab is selected. It is a `.ino` file.

Figure 19.5: The Editor page

Much of this should look familiar. `#include "thingProperties.h"` is a library that was created based on the thing chosen. The familiar `void setup()` and `void loop()` are there, as is `Serial.begin(9600)`. Notice the `ArduinoCloud.begin(ArduinoIoTPreferredConnection);`, which will connect your project to the cloud. `Void onBrightnessChange()` was set up when the thing was created to control our thing. We will come back and edit this sketch in a bit.

Click the `thingProperties.h` tab at the top to view the library for the thing that was created automatically when the property for the light was added. See Figure 19.6.

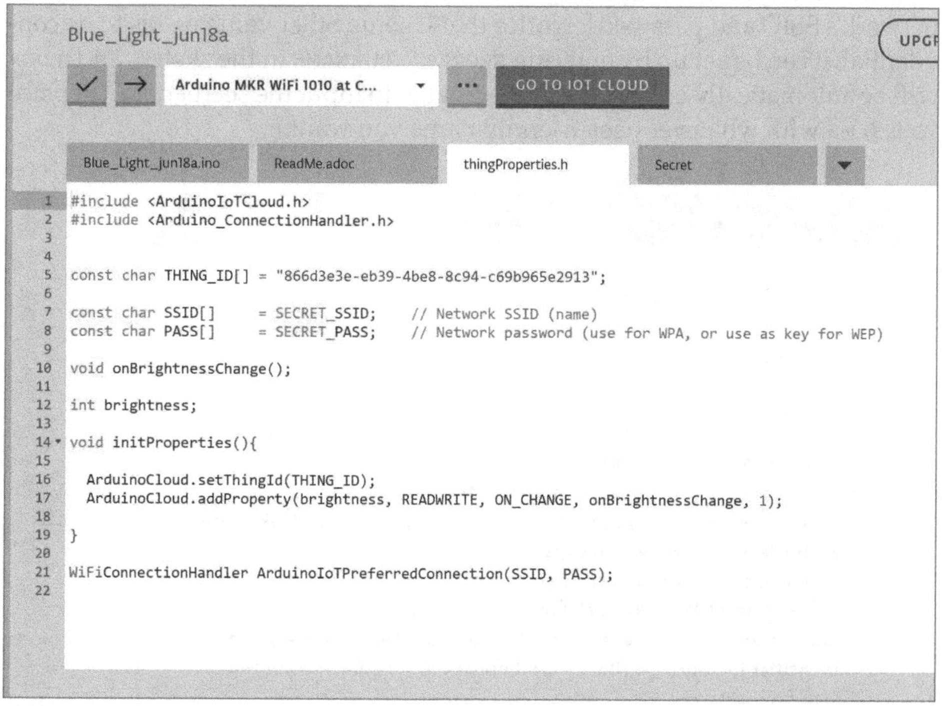

```
Blue_Light_jun18a                                                        UPGR

  ✓  →   Arduino MKR WiFi 1010 at C...  ▼   ...   GO TO IOT CLOUD


    Blue_Light_jun18a.ino    ReadMe.adoc    thingProperties.h    Secret    ▼

 1  #include <ArduinoIoTCloud.h>
 2  #include <Arduino_ConnectionHandler.h>
 3
 4
 5  const char THING_ID[] = "866d3e3e-eb39-4be8-8c94-c69b965e2913";
 6
 7  const char SSID[]     = SECRET_SSID;    // Network SSID (name)
 8  const char PASS[]     = SECRET_PASS;    // Network password (use for WPA, or use as key for WEP)
 9
10  void onBrightnessChange();
11
12  int brightness;
13
14▸ void initProperties(){
15
16     ArduinoCloud.setThingId(THING_ID);
17     ArduinoCloud.addProperty(brightness, READWRITE, ON_CHANGE, onBrightnessChange, 1);
18
19  }
20
21  WiFiConnectionHandler ArduinoIoTPreferredConnection(SSID, PASS);
22
```

Figure 19.6: The `thingProperties.h` tab

We won't be changing anything in the `thingProperties.h` file, but let's take a look at some of the code that is there. Notice the lines that say `const char SSID[] = SECRET_SSID;` and `const char PASS[] = SECRET_PASS;`. The values for `SECRET_SSID` and `SECRET_PASS` are set on the Secret tab and are the credentials used to log in to your local Wi-Fi.

At the top, `#include <ArduinoIoTCloud.h>` brings in the cloud library, which is used to sync IoT Cloud properties and sketch variables; `# include <Arduino_ConnectionHandler.h>` is the library that manages the Wi-Fi connection.

Beneath that is `const char THING _ ID []`, which holds the identification number of the thing that was assigned automatically when the thing was created. Toward the bottom of the screen is `ArduinoCloud.setThingId(THING_ID);` that tells the sketch which thing we want it to connect to. Notice that `THING_ID` is in all caps to match the constant `THING_ID` that was declared earlier.

`Void onBrightnessChange();` declares the function called whenever the `Blue_Light` (thing) property is changed by the widgets on the Dashboard, while `int brightness;` sets an integer variable named `brightness`, and `initProper-`

ties() is a function that will be called in the setup() block of the sketch and includes our thing's ID and properties.

Click the Secret tab. (See Figure 19.7.) Scroll down if you need to, so you can see the SECRET_SSID and SECRET_PASS boxes. This is where you will enter your network's SSID and password. Notice that if some other value needs to be confidential, it can be set up by entering SECRET_YOURNAME in the sketch, and a box will be automatically created on the Secret tab to input the secret data. (Replace YOURNAME with whatever user-friendly name you want.)

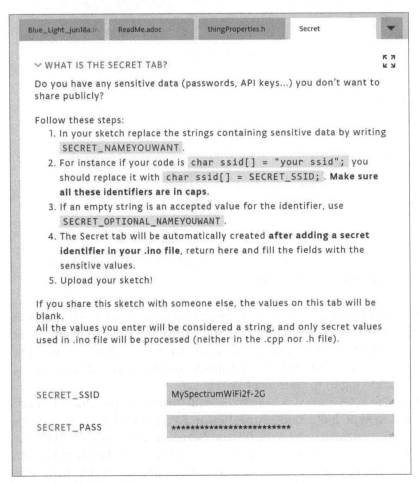

Figure 19.7: The Secret tab

Finally, click the ReadMe.adoc tab. The demo project is shown in Figure 19.8. This tab contains information about this sketch and should be edited to reflect pertinent information about the project it goes with.

Figure 19.8: The ReadMe.adoc tab

Return to `create.arduino.cc`. The screen should look like Figure 19.1. On the bottom right, click Arduino IoT Cloud. It should open showing the thing that you created earlier. See Figure 19.9.

On the menu across the top, click Dashboards. The Dashboard is the user interface that allows a user to control things and view data about things. First, click the Create Dashboard button. (See Figure 19.10.)

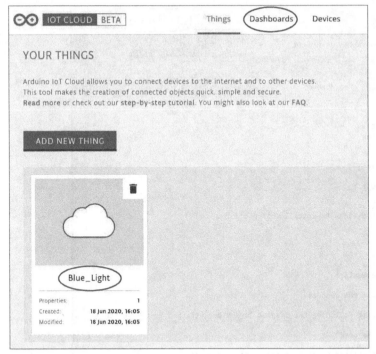

Figure 19.9: The IoT Cloud thing

Click the Add^ button, which brings up a list of widgets. Widgets are used to control or view properties of the thing in our sketch. See Figure 19.11.

The following are the available widget types:

- Switch
- Push Button
- Slider
- Stepper
- Messenger
- Color
- Dimmed light
- Colored light
- Value
- Status
- Gauge
- Percentage
- LED
- Chart

Figure 19.10: Welcome to Dashboards

For this project, we'll be using the Slider and Stepper widgets, so choose each of those. First click Add^, then Slider and Add^, and then Stepper. Your screen should now look like Figure 19.12.

1. Click inside the Slider box. A Widget Settings window should appear.

2. Change the Max value on the Value range to 255.

3. Click the Link Property button (see Figure 19.13), then click Brightness under Properties to link this widget to the brightness property of the Blue_Light thing, and finally click the Link Property button shown on the bottom right.

Figure 19.11: Widgets

4. On the next screen, click Done. The slider should now say brightness at the top and Blue_Light at the bottom.

5. Now, click inside the Stepper box and repeat steps 2 through 5.

6. Click Use Dashboard in the upper-left corner.

Now it's time to go back and edit the sketch to make it do our bidding. Perform the following steps:

1. Ensure that your Arduino board is connected to the PC's USB port.

2. Open a new browser window and navigate to create.arduino.cc.

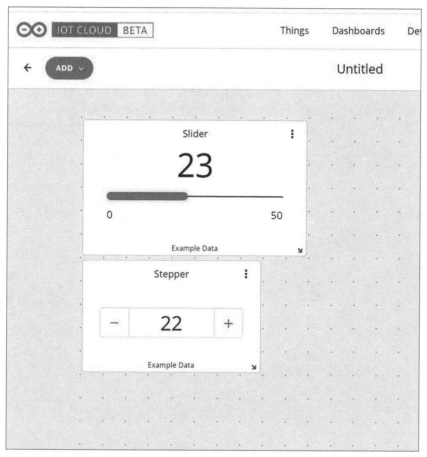

Figure 19.12: The widgets added screen

3. Click Arduino Web Editor on the top left. It should open with the sketch that you looked at earlier, with the name of your thing followed by the date and .ino.

4. If necessary, click the .ino tab.

5. Edit the sketch as shown here. The entire sketch is shown after the new entries. The new entries are as follows:

- `#define POWER A3`, which goes above `void setup()`
- `pinMode(POWER, OUTPUT);` after `void setup()`
- `int brightness = 0;` directly after the previous line
- After `void onBrightnessChange() {`, enter these two lines:
  - `analogWrite(POWER, brightness):`
  - `Serial.println(brightness);`

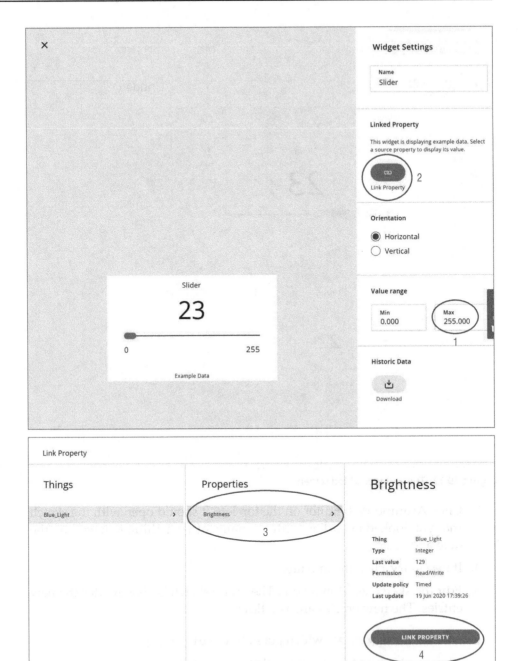

Figure 19.13: Linking the widget

Here is the entire sketch:

```
/*
  Sketch generated by the Arduino IoT Cloud Thing Blue_Light
  https://create.arduino.cc/cloud/things/866d3e3e-eb39-4be8-8c94-
c69b965e2913

  Arduino IoT Cloud Properties description

  The following variables are automatically generated and updated when
changes are made to the Thing properties

  int brightness;

  Properties which are marked as READ/WRITE in the Cloud Thing will also
have functions
  which are called when their values are changed from the Dashboard.
  These functions are generated with the Thing and added at the end of
this sketch.
*/

#include thingProperties.h
#define POWER A3  //names analog pin A3, POWER

void setup() {
  pinMode(POWER, OUTPUT); //initializes POWER pin as output.
  int brightness = 0;     //sets the LED to OFF


  // Initialize serial and wait for port to open:
  Serial.begin(9600);
  // This delay gives the chance to wait for a Serial Monitor without
blocking if none is found
  delay(1500);

  // Defined in thingProperties.h
  initProperties();

  // Connect to Arduino IoT Cloud
  ArduinoCloud.begin(ArduinoIoTPreferredConnection);

  /*
    The following function allows you to obtain more information
    related to the state of network and IoT Cloud connection and errors
    the higher number the more granular information you'll get.
```

```
        The default is 0 (only errors).
        Maximum is 4
 */
   setDebugMessageLevel(2);
   ArduinoCloud.printDebugInfo();
}

void loop() {
   ArduinoCloud.update();
   // Your code here

 }


void onBrightnessChange() {
   // Do something


analogWrite(POWER, brightness);

 Serial.println(brightness);
   }
```

6. Click the Security tab and ensure that your Wi-Fi SSID and password have been entered.

7. Click the right arrow at the top to compile, upload, and save the sketch.

8. Open the Monitor on the left of the screen to see your program at work.

9. Once you've successfully uploaded the sketch, put the Editor (Monitor) window and the Dashboard window side by side on your screen, so you can see what happens in the window as you adjust the Slider and the Stepper values. The values change in lock step, and as you adjust the values, the light appears dimmer or brighter.

Now as long as your Arduino board has power and is connected to your Wi-Fi, you can control it from anywhere through a browser!

Congratulations!

Just for Fun

"Girls, they wanna have fun. Oh, girls just want to have fun."

—Cyndi Lauper

My aunt told me once that I have a very strange idea of what fun is, but because you've read this far in this book, I'm willing to wager that your idea of fun is similar to mine. Electronics is more than work, it's fun, so here are some fun projects for you to try.

Electronic Fabrics and Wearables

This may surprise you, but a material doesn't have to be designated as "conductive" for it to be conductive. I'm sure I have some New Year's Eve tops that would fall under that category. There's enough metal in that bling to light up a holiday tree! When most people think of wearable electronics, they're thinking of either their smartwatch or those light-up holiday necklaces. The truth is that with tiny surface-mount LEDs and conductive thread, you can make anything have its own electric bling.

Figure 20.1 shows a kit from Adafruit that has five LEDs, two different switches, and a coin battery holder. Also shown is a battery, sewing needles, and a bobbin of conductive thread. Depending on where you shop, at the time of this writing, sewable LED boards can be purchased for as little as $3 U.S. for five, and a bobbin of conductive thread costs about the same. Additionally, you will need a sewing needle and a 3V battery, such as a CR2032 or CR2025. These batteries can also be purchased in packages with insulation around the battery and wire leads already attached.

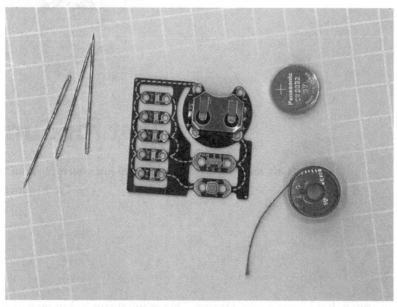

Figure 20.1: A sewable LED kit

Adafruit is not the only company that sells such tiny LEDs and kits. They can also be purchased from Sparkfun, and a quick search for *sewable LED* on Digikey will show you other companies and designs as well. For the adventurous, it's possible to make your own LED packs using surface-mount components.

Surface-mount devices (SMDs) are miniaturized versions of the components that have been used in other projects throughout this book. Figure 20.2 shows two SMDs and the packaging that they commonly arrive in.

The component on the far left of the penny is a resistor, and on the right next to it is an LED. The blue board is an SMD practice project, and it does take some practice! Tiny LEDs, resistors, capacitors, diodes, and many other components can be smaller than a grain of rice. They may be tiny, but they're used in innumerable devices now. Look closely at the LED boards in Figure 20.1. Even though the LEDs are tiny, they still function the same as larger LEDs, and the tiny circuit-limiting resistors can be seen next to the LEDs on the LED boards.

In addition, 3mm through-hole LEDs could also be used for some projects. However, the leads would need to be bent into a circle with conductive thread wrapped around them, and perhaps a bit of hot glue to hold it all in place.

Figure 20.2: SMD examples

Conductive thread generally comes in two different forms—either silver-coated thread or steel thread—and each has its advantages. The silver-coated thread is more flexible and easier to sew with, but the steel thread has less resistance. Yes, even on something this small, Ohm's law still applies. To determine how much resistance is on the thread you want to use, simply use an ohmmeter to measure it. If your silver-coated thread conductor is too long, your LEDs may be dimmer because of the thread's resistance. Remember, too, that other materials are conductive and can be used; however, like the conductive thread, they may not be insulated. It's also important to consider the lack of insulation in your design, because if one conductor crosses over another, you'll need to put an insulating material between them or they'll short out.

You'll also need to remember the rules for series and parallel circuits. If the source is only 3V, then most likely you'll need to put the LEDs in parallel with each other, as parallel branches all have the same voltage. Even tiny LEDs need at least a couple volts of electricity to light. Look again at the LED boards in

Figure 20.1. Notice that the dotted lines indicate where thread would go and that the LEDs are wired in parallel with each other.

So, now that you have these tiny LED boards and conductive thread, what can you do with them? Here are a few ideas:

- Add lights to a stuffed animal.
- Make an interactive "soft book" for a child.
- Add LEDs to a festive dress or skirt or your favorite ball cap.
- Sprinkle LEDs through a curtain.

Now let's make something.

Try This: Lighting Up a Teddy Bear

For this project, we'll light up a stuffed animal's collar. Consider creating a small project for your first one until you're comfortable with the limitations and use of conductive thread. For this project, you will need the following:

- Sewable push button
- (3) Sewable LEDs
- 3V coin-type battery
- Silver-coated conductive thread
- Sewable battery holder
- Scissors
- Regular thread of appropriate color
- Needle threader
- Sewing needles
- Stuffed animal with ribbon collar
- Hot glue and glue gun (recommended)

Consider the material where the LEDs will be mounted. Will they be mounted on top of the material or behind it? If they will be behind, make sure that they are bright enough to be seen through the fabric. You might want to use two pieces of ribbon: one for mounting the LEDs on and a sheer ribbon to cover them with.

Where will you place the battery and the switch? Where will the conductive thread need to run? Remember, there will be three lengths of thread: one for the negative side of the circuit, a second for the positive side of the circuit, and a third from the switch to the battery. At each thread hole, you'll need to make three to four loops of thread to ensure a good electrical connection. See

Figure 20.3. How long will your thread be? Will the lights be bright enough? I suggest you plan this carefully before you continue.

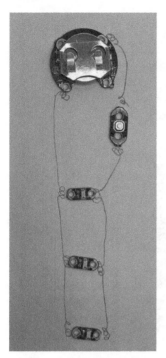

Figure 20.3: The thread circuit

Plan the spacing and orientation of the LEDs. Notice the polarity of the LEDs and the battery. Remember, the LEDs must be sewn in parallel. Mark their location.

If the LED boards are connected together on a larger board, you'll need to separate them. Cut them apart with sharp scissors or snap them apart. See Figure 20.4.

Measure how long your thread for the first section will need to be and cut a piece a bit longer, just to ensure that you have enough. Remember, you'll need to wrap the thread through each connection hole at least three to four times to make a good connection.

Fold the thread about 2 inches from one end. Press the fold together tightly and then push the thread through the eye of the needle. See Figure 20.5. Next, pull the shorter end of the thread through so that there are about 4 inches of excess. This helps to ensure that the thread doesn't fall out of the needle while you're sewing.

Figure 20.4: Separate LED boards

Now you're ready to start sewing. It's recommended that a dot of hot glue is used to hold the components in place; just be careful not to use so much that it will interfere with the conductive loops where the thread will be. Remember to check the alignment of polarity for the LEDs *before* they are glued in place. In particular, if you glue the LEDs in place, carefully checking polarity, you won't have to remember to check the polarity each time while you're sewing and can avoid having to remove the stitches later. Voice of experience here. . . .

Tie a knot in the end of the thread's longer end. You'll start sewing at one end of the circuit. The positive side of the battery is a good place to start. Push the needle up through the fabric and the connecting hole of the battery holder. Push it back down through the fabric as close to the hole as possible and pull it tight. Do this three or four times to make sure the connection is good. For the last stitch on each component, pull the thread through the loop and tighten it to make the stitches more secure. See Figure 20.6.

To go to the next connection point, either an LED or a switch, sew through the fabric at about 1/4-inch intervals so that the thread will be tight to the fabric and not loose enough to accidentally be pulled out.

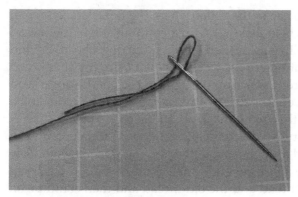

Figure 20.5: Threading the needle

The stitching at 1/4-inch intervals is called a *running stitch*. If your project allows, run the thread through the inside of the project where it won't get caught on anything. When planning your project, it's usually desirable to have stitches hidden somehow. In Figure 20.6, the stitches can be seen on the right if you look closely.

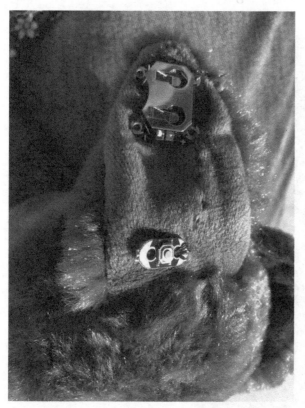

Figure 20.6: Sewing the components

The distance from the dog's ear to the collar where the LEDs are is several inches. Rather than sew the whole distance with a running stitch, the thread from the switch to the first LED was pushed through the dog's head, behind the nose and brought back out where the LED is. When bringing the thread back up from the LEDs, the thread was run through the back of the dog's head up to the ear. This was to prevent the thread from being seen and to keep the threads far enough apart that they wouldn't short out on each other.

Sometimes, the thread gets twisted and breaks. It happens. Don't panic. Get a new, straight piece of thread and go back to the last component sewn. Sew through that connection hole again several times to make a good electrical connection between the new thread and the old thread and then continue on to the next component.

On the last LED, cut the thread after sewing the last positive end. (Refer to Figure 20.3.) Start sewing again on the opposite (negative) side of the LED. When you're done, all of the positives should be sewn together, and all of the negatives should be sewn together, forming a parallel circuit.

In this example, the dog's ear was tacked back down where it started, hiding the battery and the switch. Now, a simple push of the button inside the dog's ear lights the LEDs on his collar. See Figure 20.7.

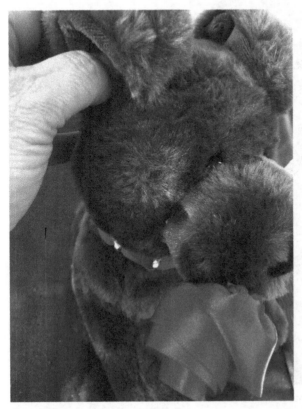

Figure 20.7: Project complete

Paper Circuits

Paper circuits are exactly what they sound like and a bit more. Using paper, cardboard, or almost anything nonconductive as a base, a circuit can be created using materials like copper tape or conductive paint, which also acts as glue. These circuits are low voltage—usually a 3V or 9V battery is used—so they are safe. For people with artistic talent, it's a great way to combine art and electronics. It's also a great way for young people to learn electronics.

The paper circuits can use regular LEDs or stick-on SMT LEDs, which are available from multiple sources on the Internet. The image on the left of Figure 20.8 shows a pack of these stick-on LEDs. Notice that they are SMT and, of course, have the current-limiting resistor built in. Both the front and back of the LED stickers have conductive copper strips for the positive and negative sides of the LED, so the sticker can be placed on top of the conductor, or the conductor can be placed on top of the sticker. The larger stick-ons on the right of Figure 20.8 are complete circuits for adding effects such as blinking, fading, twinkling, and a heartbeat flashing pattern. If you look closely, you'll see the same type of components that are used in breadboard circuits. Each circuit appears to have two transistors, two or three LEDs, and an unidentified IC.

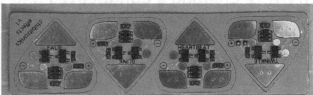

Figure 20.8: SMT stick-on LEDs and circuits

Of course, many of the other components you've learned about, such as relays, transistors, and logic chips, can also be used with paper circuits. The conductive paint or tape merely replaces wire conductors. Copper tape was found at my local hardware store, as was aluminum tape (used for ductwork), which is also conductive. Like the fiber circuits, many conductive materials that are not intended to be conductors could be used. When using these alternate materials, however, make sure that the part that touches the component's leads is conductive. Not all adhesive used on the back of these tapes is. Conductive paint can be ordered on the Internet. Yet with some experience, you can make your own using general-purpose glue and something conductive like powdered graphite.

Figure 20.9 shows conductive strips of graphite. These were made using a no. 2 pencil. The resistance was measured to give you an idea of the conductivity of the graphite. Conductivity is the inverse of resistance, so the less resistive a material is, the more conductive it is. The larger and more dense the graphite block is, the more conductive it is. Conductive paint works much the same way. The beauty of conductive paint is that, with some creativity, even large surfaces like walls could be turned into electronic circuits. Conductive paint can also be painted over (once it's thoroughly dried) so no one needs to see the conductive paint, just the finished mural or wall.

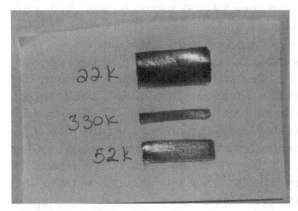

Figure 20.9: Graphite conductors

Try This: Creating a Conductive Paint Circuit

This simple project will give you the feel of working with conductive paint. For this project, you will need the following:

- Conductive paint
- 9V battery
- LED
- Paper
- Pencil

Determine your circuit. The image shown in Figure 20.10 is a simple series circuit. Note that as constructed in the image, a current-limiting resistor isn't needed because there is sufficient resistance in the conductive paint.

1. On your paper, mark the distance between the positive (+) and negative (-) battery terminals.

2. Determine where the painted conductors will be located and draw them on the paper.

3. Fill in the drawn lines with conductive paint, putting larger pads of paint where the LED will be.

4. Note the positive and negative sides of the LED and add it to the circuit by placing a lead of the LED in each pad.

5. Place a generous dot of conductive paint on top of the LED leads. The paint acts as glue, in addition to being conductive.

6. Allow the paint to dry completely! The paint is not conductive until it is dry.

7. Place the battery on the circuit with the proper polarity for the LED to light.

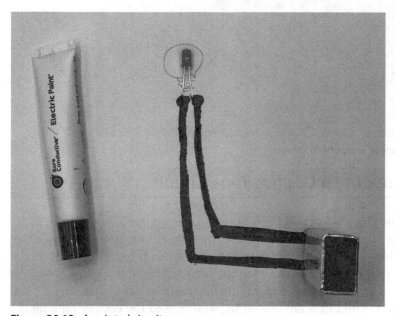

Figure 20.10: A painted circuit

The conductive paint can be painted over with acrylic paint to cover the conductive strips with color. This could be done to light up paper flowers or almost anything, really. A paper switch could run a relay, which in turn would run a fan or a 110V lamp. The circuits are like any other, limited only by your imagination. Figure 20.11 covers a painted circuit with acrylic paint. You can't see the conductors, but they are there.

What will you do with conductive paint? Paint a light switch on a wall? Control a paper lamp? Create an interactive wall map? Create a proximity sensor? Let your imagination run wild.

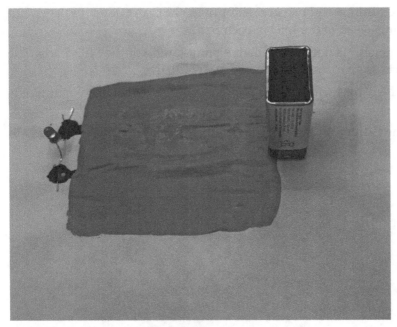

Figure 20.11: Paint over a painted circuit

Try This: Creating a Copper Tape Circuit

This simple circuit can be used to light up a greeting card. See Figure 20.12. The tape strips on the left side of the page come into contact with the positive side of the battery when the paper is folded, and when the user presses on the "Press Here" spot on the front (beneath the pressing fingers, Figure 20.13), the conductive tape strips make contact with the positive side of the circuit just above the battery. The LEDs are placed in parallel, and the copper tape on the negative side of the circuit runs under the 3V battery to its negative side. The battery is secured in place with cellophane tape, but be certain to tape just the edges, not the center of the battery, so it can still make the connection when the page is folded. Do not glue the battery to the page unless the glue is conductive; otherwise, it would prevent the negative side of the battery from making a connection with the negative side of the circuit.

Paper circuits can become interactive by adding sliding parts or spacers that act as switches or by using the special feature circuit stickers and LEDs mentioned previously. Figure 20.13 shows the completed circuit.

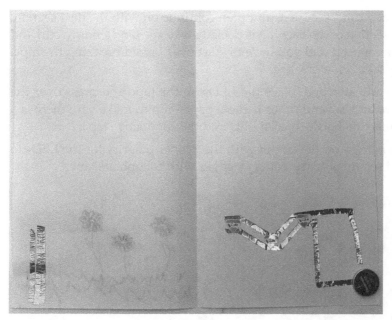

Figure 20.12: A copper tape circuit

For this project, you will need the following:

- 12 inches of copper tape
- 3V battery
- (3) Stick-on LEDs
- Paper
- Pencil

Optional materials include the following:

- Glue stick
- Sheet of cardstock

1. Determine the placement of the stick-on LEDs and either mark the spot or attach the LEDs to the paper, making sure they are oriented properly for polarity.

2. On your paper, draw lines to indicate where the edge of the conductive tape needs to be, making sure that the strips of tape will be close enough to make contact with each side of the LEDs and that the positive and negative sides of the circuit will make a solid connection with the battery.

3. Peel back the first 1/4 inch of conductive tape and secure it in place. Continue to peel away only a small portion of the paper backing from the conduc-

tive tape at a time, secure it, and then continue on with the next section. If a large section of the backing is peeled away, the tape tends to curl and stick back on itself and you have to start over. Avoid breaking the tape. This may take some practice.

4. If not already attached, place the LEDs on the tape and press over the conductive area several times using a paper clip to make sure there is a solid connection between each LED and the conductive tape.

5. Place the battery on the circuit with the proper polarity for the LEDs to light, and secure only the battery's edges with cellophane tape.

Figure 20.13: The copper tape card front

To hide the circuit within a card, use a glue stick along the sides of the inside, as shown in Figure 20.14, but don't glue close to the "Press Here" spot. Then glue the back of the folded card onto the front of a piece of half-folded card stock.

Try This: Building Squishy Circuits

Squishy circuits are made using conductive dough and nonconductive dough or other nonconductive material as insulators. They use either 3V or 9V DC power, so they're a safe and fun way for kids to learn about electricity.

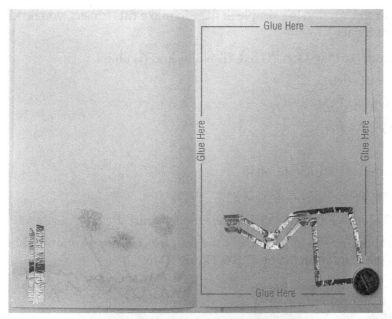

Figure 20.14: Finishing the card

You can use the commercial product Play-Doh, which is conductive, or you can make your own conductive and nonconductive dough with ingredients you likely already have on hand in your kitchen. Recipes for both conductive and nonconductive clay can be found in several places online, but the one I suggest is squishycircuits.com/pages/dough-recipes.

It is important that an insulating material be placed between the positive side of the circuit and the negative side of the circuit so that the electrons will be forced to go through the load—whether it is an LED, buzzer, or anything else—instead of simply shorting out through the dough. The main difference between most conductive and nonconductive dough is that the conductive dough contains salt, and the nonconductive dough contains sugar. Also, be sure to use distilled water if making nonconductive dough instead of tap water. Tap water often contains traces of minerals that will make the water conductive. When making the dough, I found that I needed a bit more flour than the recipe calls for, so be sure you have a bit extra on hand.

The project for this section is a simple lit flag. To make this project, you need the following materials:

- (3) Containers of Play-Doh, (1) red, (1) white, and (1) blue
- Wooden craft stick
- (8) 3mm white LEDs
- 9V battery with snap
- Scissors
- Small snap-blade knife (optional)
- Cellophane tape

1. Cut four 1/2-inch long pieces from the wooden craft stick.
2. Use the cellophane tape to connect the four pieces into a square. This square is the insulating material. See Figure 20.15.

Figure 20.15: The insulating square

3. Fill the inside of the square with blue Play-Doh.
4. Roll out some blue Play-Doh into a log and flatten it to about 1/4-inch thick, cut to the height of the craft-stick square.
5. Wrap the square with the blue Play-Doh and smooth the edges together.
6. Ensure that the Play-Doh inside and outside of the wooden square does not touch. Scrape any excess away with scissors or the snap-blade knife if needed.
7. Place the desired number of white LEDs with their positive leads sticking into the center of the blue square and their negative leads sticking into the outside of the blue square, separated by the craft wood.
8. Insert the positive lead from the battery into the bottom of the center blue square and the negative lead into the outside of the blue square. The LEDs should light. See Figure 20.16.

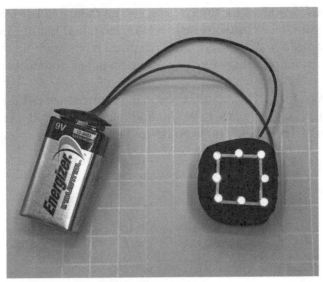

Figure 20.16: The lit blue dough square

9. If desired, form the red and white Play-Doh into stripes for the flag, as in Figure 20.17.

Figure 20.17: The lit flag

Now that you have a few ideas for fun circuits, get out there and be creative! Spend some time in your local hardware store where you'll find lots of conductive materials. Instead of discarding shoe boxes or pizza boxes, turn them into electronic pieces of art using things like paper clip switches and conductive screws.

This chapter has been fun, and whether you're in it for fun or serious business, working in electronics is a great place to be. Take a look at Chapter 21, "What's Next?", for resources that will help you keep up-to-date with this exciting and growing field.

What's Next?

"Life could be a dream... Sh-boom sh-boom..."

—James C. Keyes, Carl Feaster, Floyd McRae, Claude Feaster, James W. Edwards

You now know so much more about electronics than the average person that the question is, "What can you dream of and turn into a reality with that knowledge?"

The World Is Your Oyster

The Internet of Things (IoT) is growing by leaps and bounds with sensors everywhere. We have security systems, thermostats, and lights that can be monitored or manipulated remotely with our cell phones. Even the air filter on my furnace is connected to the Internet via Wi-Fi. Every day, people are discovering new ways to use IoT to make life more comfortable or businesses more profitable, or using IoT devices simply for fun. It may be the air quality in the building you work in that needs to be monitored and automatically adjusted, or an assembly line in a manufacturing plant that needs greater efficiency. Perhaps someone could develop an IoT device to tell us when the surf is up! (Last summer in Cape Cod, we used an app to show us where the shark sightings were.) Some places use sensors to indicate where the vacant parking spots are, which helps mitigate the pollution and frustration caused by cars endlessly circling large parking lots. IoT sensors are used in systems that turn solar panels to face the sun as it moves across the sky. Someone I know even uses IoT collars and a smart cat door to monitor whether their cats are inside or out.

Whatever your passion is, potential and opportunity exist for you to solve a problem or question using IoT. Perhaps my favorite fun idea is the pet-monitoring webcam that lets users dispense treats to their furry friends from a remote location using their cell phones. The bottom line is, IoT is limited only by your imagination and education, and you've already taken the first step by learning the information in this book. If you're an enthusiast wanting to create quirky tools for your home or a serious inventor, there is room for you in IoT, and you don't have to go it alone. A quick look at your favorite job bank can reveal hundreds (or thousands, depending on your search area) of IoT-related jobs, such as senior embedded firmware engineer, IoT sales, data analytics, and so on.

Life is but a dream. . .dream it up and create it!

Recommended Reading and Resources

Electricity is the same as it was when Benjamin Franklin shocked himself with it in the 1700s. What has changed, and is evolving daily, is the way that we use it to manipulate our world. Every day, new devices are created for basic electronic circuits, digital electronics, IoT sensors, and such. This means that it's important to stay current with what's going on in the world of electronics. In this field, you can't simply learn it once and be done.

One of the best places to find out what's new with electronics and electricity, or anything really, is to watch the manufacturers' websites and distributors' new items. Many of them have mailing lists that you can sign up for to easily receive new information without searching daily for it. Over time, you'll find your own favorite resources, but for now I'll share just a few of mine with you.

- Ti.com is the site for Texas Instruments (TI), which manufactures a plethora of electronics devices. Click the Products menu, and you'll see many of the components talked about in this book and more. The Support & Training menu has online tutorials, technical documentation, and precision labs that include videos and PDFs teaching you how to use TI products.

- Digikey.com is a major distributor of electronic components. Its website also provides online resources that can help you design a project, choose parts, and find articles from manufacturers and makers that can give you ideas about projects, or solutions to problems you may be struggling with. Their Scheme-It program is an easy way to design circuit schematics, and it links to their parts database so that when the design is done, a bill of materials (BOM) can be printed. Hover over Tools on the menu bar to see what other resources are available.

- `Mouser.com` is another favorite source of parts. Click Technical Resources on the menu bar to find articles and links to eBooks from manufacturers.

- `Raspberrypi.org` is a great place to start if you want to work with that platform. What is Raspberry Pi, you ask? The Raspberry Pi looks like an Arduino, but instead it's a tiny computer. On the website, you can find forums, projects, blogs, coding tutorials, and a plethora of other information. Considering its computing power in such a small package, a Raspberry Pi could become part of your IoT solution.

- `Arduino.cc`, of course, is the best place to learn about Arduino and its capabilities. Click Resources and then Reference to find the nitty-gritty of the Arduino language.

- My favorite website, of course, is `cliffjumperTEK.com`. Look there for basic electronics tutorials and videos of some of the projects in this book along with others.

Words of Encouragement

Virtually anyone can do electronics. From the conductive play clay for your kids to corporate IoT solutions, it's the same process.

1. Begin with an idea.
2. Research/gain knowledge.
3. Do the math!
4. Test it.
5. Do it!

It's also perfectly acceptable and advisable to ask someone else when you're not sure about something, especially if safety is involved. Cultivating friendships with others in the electronics field is advisable. It's always more fun to talk with someone who understands what you're talking about, instead of the blank stares you get when you chatter excitedly about pulse width modulation or your favorite oscilloscope with the uninitiated.

Ignore the people who discourage your creativity. Had I listened when people said, "Girls don't do that!" where would I be today? Certainly not writing this book about electronics for you. I'm not here to tell you it's always easy. It isn't. As the "only girl in the room," I had to work harder and be assertive to prove my mettle in a male-dominated field. We all have our challenges to overcome; mine happened to be my gender in a time when people thought women belonged in the kitchen, not the boardroom, or, in my case, the garage. Luckily, things are different now than they were 30 years ago. I'm so encouraged when I see people, regardless of their gender, age or whatever other factor the world chooses to identify them with, doing what they feel passionate about, ignoring the naysayers, and making the world better for it.

We all have gifts, you included, and if you're like me and yours involves electricity, solder, sensors, buzzers, and such, then go for it. Share your gifts with the world. There's a lot of room for everyone in this field, and the world needs your unique perspective, talents, and creativity.

Dream big. Do great things. You are awesome.

Index

A

AC (alternating current), 56, 342–343
AC power source, schematic symbol for, 43
accuracy, of sensors, 6
acid, 308
active filtering, 350
actuators, 44
Adafruit, 405–406
ADC (analog-to-digital converter), 61
adding
  station tuning to Arduino FM radio, 377–381
  switches, 48–52
  switches to circuits, 74–78
Ah (ampere-hours), 125
aiding sources, 153–154
air core inductors, 225, 226, 329
allotropes, 17
alternating current (AC), 56, 342–343
aluminum, as a conductor, 16
AM (amplitude modulation), 369, 371
American National Standards Institute (ANSI), 41–42
American Society of Mechanical Engineers (ASME), 41–42
American Wire Gauge (AWG), 151, 152
ammeters (Arduino), 102–107
ampere-hours (Ah), 125
amperes, 18
amplifiers
  operational, 272
  using transistors as, 187–191
amplitude, 342
amplitude modulation (AM), 369, 371
analog pins (Arduino Uno), 59
analog signal, digital signal *vs.*, 60–62

analog-to-digital converter (ADC), 61
`analogWrite` function, 301–303, 323
AND gates, 277–280
angle of declination, 219
angle of variation, 219
anode
  defined, 69
  determining, 171–172
ANSI (American National Standards Institute), 41–42
antennas, schematic symbol for, 43
arching, 47
Arduino
  about, 53–54
  adding switches to circuits, 74–78
  ammeters, 102–107
  analog *vs.* digital, 60–62
  Arduino board, 54–59
  Arduino IDE, 62–65
  changing pins, 71–72
  continuity testers, 107–108
  creating Arduino running lights, 72–74
  creating Arduino-controlled circuits, 65–71
  creating dimmable camp lights, 109–115
  creating FM radios, 371–381
  creating laser security systems, 255–262
  creating running lights, 72–74
  displaying PV output on, 315–323
  environment, 54
  forum, 390
  measuring electricity using, 95
  ohmmeters, 100–102
  PWM and, 299–305
  shields, 54
  using serial monitors, 78–81

427